Kurt Brauchli

Telemedicine for Improving Access to Health Care in Low-Resource Areas

Kurt Brauchli

Telemedicine for Improving Access to Health Care in Low-Resource Areas

From Individual Diagnosis to Strengthening Health Systems

Südwestdeutscher Verlag für Hochschulschriften

Impressum/Imprint (nur für Deutschland/ only for Germany)
Bibliografische Information der Deutschen Nationalbibliothek: Die Deutsche Nationalbibliothek
verzeichnet diese Publikation in der Deutschen Nationalbibliografie; detaillierte bibliografische
Daten sind im Internet über http://dnb.d-nb.de abrufbar.
Alle in diesem Buch genannten Marken und Produktnamen unterliegen warenzeichen-, marken-
oder patentrechtlichem Schutz bzw. sind Warenzeichen oder eingetragene Warenzeichen der
jeweiligen Inhaber. Die Wiedergabe von Marken, Produktnamen, Gebrauchsnamen,
Handelsnamen, Warenbezeichnungen u.s.w. in diesem Werk berechtigt auch ohne besondere
Kennzeichnung nicht zu der Annahme, dass solche Namen im Sinne der Warenzeichen- und
Markenschutzgesetzgebung als frei zu betrachten wären und daher von jedermann benutzt
werden dürften.

Verlag: Südwestdeutscher Verlag für Hochschulschriften Aktiengesellschaft & Co. KG
Dudweiler Landstr. 99, 66123 Saarbrücken, Deutschland
Telefon +49 681 37 20 271-1, Telefax +49 681 37 20 271-0, Email: info@svh-verlag.de
Zugl.: Basel, Universität, Diss., 2006

Herstellung in Deutschland:
Schaltungsdienst Lange o.H.G., Berlin
Books on Demand GmbH, Norderstedt
Reha GmbH, Saarbrücken
Amazon Distribution GmbH, Leipzig
ISBN: 978-3-8381-0390-7

Imprint (only for USA, GB)
Bibliographic information published by the Deutsche Nationalbibliothek: The Deutsche
Nationalbibliothek lists this publication in the Deutsche Nationalbibliografie; detailed
bibliographic data are available in the Internet at http://dnb.d-nb.de.
Any brand names and product names mentioned in this book are subject to trademark, brand or
patent protection and are trademarks or registered trademarks of their respective holders. The
use of brand names, product names, common names, trade names, product descriptions etc.
even without a particular marking in this works is in no way to be construed to mean that such
names may be regarded as unrestricted in respect of trademark and brand protection legislation
and could thus be used by anyone.

Publisher:
Südwestdeutscher Verlag für Hochschulschriften Aktiengesellschaft & Co. KG
Dudweiler Landstr. 99, 66123 Saarbrücken, Germany
Phone +49 681 37 20 271-1, Fax +49 681 37 20 271-0, Email: info@svh-verlag.de

Copyright © 2009 by the author and Südwestdeutscher Verlag für Hochschulschriften
Aktiengesellschaft & Co. KG and licensors
All rights reserved. Saarbrücken 2009

Printed in the U.S.A.
Printed in the U.K. by (see last page)
ISBN: 978-3-8381-0390-7

Contents

Foreword   xv

Acknowledgement   xvii

Summary   xxi

I. Overview   1

1. Introduction   3

2. Goals and Objectives   7
   2.1. Overall Goals 7
   2.2. Objectives 7

3. Definitions and Context   9
   3.1. Information and Communication Technologies in Medicine 9
   3.2. Telemedicine 10
       3.2.1. Definitions 10
       3.2.2. Technology 12
   3.3. Specific Aspects for Resource-Constrained Areas 13
       3.3.1. Potential Benefits 14
       3.3.2. Limitations 14

4. Literature Overview   17
   4.1. Evolution of Telemedicine 17
   4.2. Telemedicine in Developing Countries 17
       4.2.1. Impact of Telemedicine in Developing Countries 19
   4.3. Medical Specialties 20
       4.3.1. Radiology 20
       4.3.2. Pathology 22
              4.3.2.1. Telepathology Systems 23
              4.3.2.2. Cytology and Hematology 24
              4.3.2.3. Telepathology in Developing Countries 24
       4.3.3. Dermatology 25

i

4.3.4.	Other Disciplines	25
4.3.5.	Telehealth or e-Health	26

II. The iPath Telemedicine System   27

5. Description of the iPath Telemedicine Platform   31
- 5.1. Technical Background of the iPath-Server 31
  - 5.1.1. Basic Functionality of iPath 31
  - 5.1.2. Application Layers 32
  - 5.1.3. The Data Model 32
  - 5.1.4. Visualisation 34
  - 5.1.5. Modular Design 36
  - 5.1.6. Security 36
  - 5.1.7. Advanced Functionality 37
- 5.2. Distribution of iPath 41
  - 5.2.1. Requirements for Installation of an iPath-Server 41
  - 5.2.2. Helper Applications 42
  - 5.2.3. Licensing and Availability 42

6. Telemicroscopy by the Internet Revisited   43
- 6.1. Introduction 44
- 6.2. Using the Internet for Telemedical Applications 44
- 6.3. A Way Out: Distributed System for Telemicroscopy 46
  - 6.3.1. Connecting the Microscope: the Microscope Control Program 46
  - 6.3.2. Using the Microscope at a Distance: the Client Web Page 47
  - 6.3.3. The Telepathology Server 48
- 6.4. Discussion 49
  - 6.4.1. An Open Standard 49
  - 6.4.2. Use for Frozen Section Diagnosis 49
  - 6.4.3. Combining Dynamic Telemicroscopy with 'Store and Forward' 50
  - 6.4.4. Future Directions 50
- 6.5. Conclusion 51

7. Applications of the iPath Server at the University of Basel   53
- 7.1. The iPath-Server at the University of Basel 53
- 7.2. Overview of Applications and Groups 55
  - 7.2.1. Groups with over 100 Cases 55
  - 7.2.2. Other Interesting Groups 61
  - 7.2.3. Virtual Portals 62
- 7.3. iPath in Developing Countries 64
- 7.4. Other Applications of iPath Software 65

8. iPath – Telemedicine Platform for Low Resource Settings   **69**
  8.1. Introduction 70
  8.2. iPath - a Hybrid Web and Email Based Telemedicine Platform 70
      8.2.1. Telemedicine Platform at the University of Basel 72
  8.3. Case Studies 74
      8.3.1. Teledermatology in Port St. Johns, South Africa 74
      8.3.2. Telepathology on Solomon Islands 75
  8.4. Discussion 76

III. Case Studies   77

9. Telepathology on the Solomon Islands–two years experience   **79**
  9.1. Introduction 80
  9.2. Methods 80
  9.3. Results 81
  9.4. Discussion 83

10. Diagnostic Accuracy of Telepathology on Solomon Islands   **87**
  10.1. Introduction 87
  10.2. Material & Methods 88
      10.2.1. Telepathology System 88
      10.2.2. Review 89
  10.3. Results 90
      10.3.1. Material Submitted 90
      10.3.2. Diagnostic Discrepancies 92
      10.3.3. Problematic Cases 93
      10.3.4. Reasons for Diagnostic Discrepancies 93
      10.3.5. Development Over Time 95
  10.4. Discussion 97
      10.4.1. Telepathology and Continuous Medical Education 99
      10.4.2. Beyond (Tele-)pathology 99
  10.5. Conclusions 100

11. Telepathology at the Sihanouk Center of Hope, Cambodia   **103**
  11.1. Introduction 103
  11.2. Material & Methods 104
      11.2.1. Laboratory 104
      11.2.2. Telepathology 105
  11.3. Results 105
      11.3.1. Diagnostic Discrepancies 106
      11.3.2. Reasons for Diagnostic Discrepancies 109
  11.4. Discussion 109

   11.4.1. Way Forward 111

12. Teledermatology in the Transkei   **113**
 12.1. Introduction 113
 12.2. Material and Methods 114
   12.2.1. Port St. Johns 114
   12.2.2. Tsilitwa 115
   12.2.3. The UNITRA Telemedicine Network 116
 12.3. Results 117
 12.4. Discussion 119

13. ICT for Distant Medical Collaboration in Ukraine   **123**
 13.1. Introduction 124
 13.2. Materials and Methods 125
 13.3. Results 127
   13.3.1. User statistics 127
   13.3.2. Case statistics 128
   13.3.3. Comment statistics 128
 13.4. Clinical Application 129
 13.5. Discussion 131
 13.6. Conclusion 133

IV. Consolidation of Results and Discussion   **135**

14. Telemedicine in Resource-Constrained Areas   **139**
 14.1. Telepathology 139
   14.1.1. The Role of the Referring Physician's Clinical Background 140
   14.1.2. Extending Existing Collaboration via Telemedicine 141
   14.1.3. Educational Impact 142
 14.2. Regional Networks 142

15. Telemedicine and Health Systems   **145**
 15.1. Potential Benefits 146
   15.1.1. Purpose of Information Exchange 146
   15.1.2. Partners in Information Exchange 148
 15.2. Acceptance 152
 15.3. Cost Effectiveness 153
 15.4. Deployment Strategies 154
   15.4.1. Selection of Technology 154
   15.4.2. The Process of Implementation 156
   15.4.3. Decentralised Collaborative Networks 159
 15.5. Towards a Grid of Decentralised Networks 162

15.6. Recommendations 163

16. Ongoing Projects 165
16.1. Educational Impact of Telepathology Consultations 165
16.2. Telecytology Study at NRH 165
16.3. Solomons Islands Telemedicine Network 166
16.4. Uzbekistan Telemedicine Network Project 169

17. Conclusions 171

Bibliography 175

List of Figures

5.1. Application layers and interfaces of iPath 33
5.2. Organisation of a virtual institute 38
5.3. Distributed presentations with iPath 40

6.1. Network layout for telemicroscopy 45
6.2. The telemicroscopy client application 47
6.3. Archived telemicroscopy session 48

7.1. Monthly user sessions 54
7.2. Monthly case submissions 55

8.1. A typical case in iPath 71
8.2. Consultations from developing countries 73
8.3. Size of images submitted to iPath 73

9.1. Web-interface of iPath 82

10.1. Age distribution of patients at NRH 91
10.2. Grade of TP diagnosis 93
10.3. Development of diagnostic discrepancies over time 95
10.4. Development of quality of image selection over time 97

11.1. Age distribution of patients from SHCH 106

12.1. Tsilitwa clinic 116
12.2. Results of teledermatology consultations in Port St. Johns 119
12.3. Case submission from Tsilitwa 120

13.1. Example of a consultation from Ukraine 126

15.1. Purpose of information exchange 147

16.1. Solomon Islands National Telemedicine Network 168

List of Tables

4.1.	Literature search	21
7.1.	Usage statistics of the iPath-server at the University of Basel	54
8.1.	Usage statistics of iPath in December 2004	72
8.2.	Telepathology consultations from Honiara	76
9.1.	Telepathology consultations from the National Referral Hospital in Honiara	82
9.2.	Results of the virtual institute	83
10.1.	Topography of submitted material submitted from NRH.	91
10.2.	Diagnostic discrepancies vs. class of disease	92
10.3.	Problematic cases submitted from Honiara.	94
10.4.	Reasons for discrepant diagnosis.	96
11.1.	Distribution of diagnostic discrepancies at SHCH	107
11.2.	Problematic consultations from SHCH	108
11.3.	Reasons for discrepant diagnosis .	110
12.1.	Diagnoses made by teledermatology.	118
13.1.	Distribution of material	128
13.2.	User activity in the USPHP groups	129
14.1.	Diagnostic accuracy of telepathology	139
14.2.	Factors influencing diagnostic accuracy of telepathology	140

List of Acronyms

ADSL .... Asynchronous Digital Suscriber Line

AIDS .... Acquired Immunodeficiency Syndrome

AJAX .... Asynchronous Javascript and XML

API .... Application Programming Interface

ARV .... Antiretroviral drug
medications for the treatment of infection by retroviruses, primarily HIV

ATA .... American Telemedicine Association

ATM .... Asynchronous Transfer Mode

CGI .... Common Gateway Interface

CME .... Continuous Medical Education

CMS .... Content Management System

CRT .... Cathode ray tube (display)

CT .... Computer Tomography

DHIS .... District Health Information System

DICOM .... Digital Imaging and Communication in Medicine
standard format for storage and exchange of digital images in medicine. Mainly used in radiology

DR .... Digital Radiology

DSL .... Digital Subscriber Line
family of protocols for high speed trasnmission, often over telephone lines

ECG .... Electocardiogram

EHR .... Electronic Health Record

EMR .... Electronic Medical Record

EU	European Union
FNA	Fine Needle Aspiration
FOSS	Free and Open Source Software
FTP	File Transfer Protocol
GP	General Practitioner
GPL	General Public License open-source software license developed by the Free Software Foundation
GSM	Global System for Mobile Communications the most popular standard for mobile phones
HDD	Harddisk drive
HE	Hematoxylin Eosin
HIV	Human Immunodeficiency Virus cause of the disease known as AIDS
HTML	Hypertext Markup Languagexs
HTTP	HyperText Trasfer Protocol
ICTs	Information and Communication Technologies
IHC	Immunohistochemistry an assay to detect antigens in tissue by the use of enzyme or fluorescence linked antibodies
ISDN	Integrated Sevices Digital Network
JAVA	Java programming language a platform independent programming language developed by Sun Inc. (http://java.sun.com)
JPEG	Joint Photography Expert Group file format for compressing digital images. JPEG is a lossy compresseion, especially suitable for photographic images
MeSH	Medical Subject Headings
MP3	MPEG-1 Audio Layer 3 a common audio compression algorithm
MRI	Megnetic Ressonace Imaging

NGOs	Non Governmental Organisations
NRH	National Referral Hospital
NST	Norwegian Centre for Telemedicine
PDF	Portable Document Format document format developed by Adobe Inc. for exchange of (printable) documents between different computer systems
PGP	Pretty Good Privacy public key based encryption standard for emails and computer documents
PHP	PHP Hypertext Processor popular programming language for development of web applications. (http://www.php.net)
RAFT	Resseau de l'Afrique Francophone pour la Telemedicine
RDBMS	Relational Database Management System
SHCH	Sihanouk Hospital Centre of Hope
SMS	Short Message Service a form of text messaging on cell phones
SMTP	Simple Mail Transfer Protocol the simple mail transport protocol regulates what goes on between the mail servers
TP	TelePathology
UNDP	United Nations Development Programme
UNITRA	University of Transkei former University of Transkei which is now part of the new Walter Sisulu University of Science and technology (WSU)
USPHP	Ukraine Swiss Perinatal Health Programme
UTF-8	Unicode Transformation Format 8 ISO standard for encoding multi-language text documents and supports of all charatcter types (e.g. cyrillic, georgina, chinese, etc.)
VC	video conferencing
VFW	Video for Windows
VGA	Video Graphics Array an analog computer display standard with 640x480 pixels resolution.

VHF	Very High Frequency a reserved frequency band for long distance radio communication
VIRIN	Virtual Institute
WHO	World Health Organisation
XGA	eXtended Graphics Array computer display standard best known as a synonym for the 1024x768 pixels resolution
XML	eXtensible Markup Language
XMLRPC	XML encoded Remote Procedure Call
XSL	eXtensible Stylesheet Language

Foreword

"Why don't you come to Solomon Islands and try to set up a telepathology link for us?", Hermann Oberli asked me in 2001. Before I really knew where these islands were, I had agreed to try to help him set up a telepathology service for the National Referral Hospital in Honiara, the country's capital, where he was working as a surgeon at that time. The possibility to deliver pathology diagnosis over Internet to a place as remote as Honiara added a fascinating dimension to our telepathology project, where so far we had considered the other end of Switzerland a far away place.

However, once the technical problems were solved, new questions came up: How accurate are diagnoses delivered over telepathology under the given conditions? How useful are such distance diagnoses for the surgeons? Despite these uncertainties the idea spread rapidly; and soon there were consultations coming from Cambodia, Bangladesh, Laos, Iran and other countries. How should we organise our group of volunteering pathologists to address the increasing number of consultations?

Yet another dimension was added to the project when we were approached by people who wanted to use the same platform for the organisation of regional telemedicine networks. This raised the question whether this form of telemedicine was only feasible for international collaborations or whether regional networks within resource-constrained areas were equally feasible. By a twist of fate I got in contact with Lech Banach from the University of Transkei in Mthatha, South Africa, and the idea for a project within one of the least developed regions of South Africa emerged. At this time I decided to turn the whole thing into a PhD project; and in the end I spent a most interesting year as a guest researcher in Mthatha trying to implement a regional telemedicine network.

As a result of these various collaborations with different countries and within different medical specialties and on all levels of health systems we started gradually gaining a better insight into health care systems of developing countries. While there had been several pilot projects on implementing telemedicine services in developing countries, we came to realise that there was little understanding on the impact of telemedicine on health care systems in resource-constrained areas: Can telemedicine improve access and quality of care or decrease cost of care delivery? Will telemedicine only help to improve diagnosis for an individual patient or would it be possible to use telemedicine as a vector for a sustainable knowledge transfer and thus contribute to strengthening health systems at large?

The research presented in this thesis is not based on an initial hypothesis but has been driven by very practical needs. This has led to a series of actions from which new questions arose, each demanding its own reflection. Although the development of a technical platform feasible for organising telemedical collaboration within a large number of participants working under very different conditions could be seen as an innovation of its own, the aim of this thesis is to achieve a better understanding of how telemedicine impacts on health systems, to consolidate our experiences from various different projects and to analyse the possibilities of telemedicine to strengthen health systems.

Acknowledgement

The presented thesis is a result of many collaborations that were established between the "iPath" telemedicine project at the Department of Pathology in Basel and various institutions and individuals around the globe – many of them from developing countries. These include in particular the former University of Transkei (now Walter Sisulu University), where I spent one year as a guest student, the National Referral Hospital (NRH) on Solomon Islands, the Sihanouk Centre of Hope (SCHC) in Cambodia, the Swiss Centre of International Health at the Swiss Tropical Institute and many others. Numerous people were involved in many ways in the realisation of this work – their help is most greatly acknowledged.

My sincerest thanks are addressed to Prof. Martin Oberholzer (Dept. of Pathology, University of Basel), who has initiated the whole project and who has always been ready to discuss new ideas and to test new software elements and who has always offered the necessary support and motivation to pursue this work. I feel indebted to Prof. Lech Banach (Department of Pathology and Telemedicine Unit, Walter Sisulu University, Mthatha, South Africa), who has enabled my stay in South Africa and organised the two conferences on telemedicine in the Eastern Cape and who, together with his wife Barbra, has made South Africa our home away from home. Equally grateful I am to Prof. Marcel Tanner (Swiss Tropical Institute, University of Basel) for accepting to be the faculty representative for this thesis and for pushing towards a broader view of telemedicine to strengthen health care systems and to Prof. Antoine Geissbuhler from the University Hospital of Geneva for accepting to be the external referee.

The whole project would probably never have started without enthusiasm and commitment of Dr. Hermann Oberli (then Head of Orthopaedic Surgery at NRH). His vision of a pathology service on Solomon Islands by means of telepathology finally enabled my visit on Solomon Islands in 2001 to set up the histology lab and a telepathology workstation, which was my first exposure to health care under severely resource-constrained conditions. This whole endeavour was only possible with the help of the fabulous team in Honiara: In particular Michael, Anna, Wilson and Andrew of the medical laboratory. Greatest thanks must also go to Dr. Rooney Jagilly, the surgeon, who has photographed and submitted all the specimen prepared by the lab. The friendship and hospitality that we experienced on Solomon Islands was simply outstanding.

My thanks are addressed to Dr. Gerhard Stauch and Dr. Chhut Serey Vathana, who established the telepathology service at the SHCH, from where over 1400 tele-consultations have originated over the past four years: Their project provided by far the largest amount of data evaluated in this study.

I am very grateful to all the people contributing to the tele-dermatology project in South Africa, particularly to Dr. Don O'Mahony, Dr. Louise Lagrange, Patricia Madikane, Chris Morris; and to the team from the Swiss Centre for International Health and their Ukrainian partners, especially to Marc Blunier, Martin Raab, Dr. Andrey Solodarenko and Dr. Anton Vladzymyrskyy.

The telepathology projects evaluated during this work would not have been possible without the help of a large pathologists' team who voluntarily accepted consultations from the various partners in developing countries. In particular I would like to thank Prof. Kunze from Dresden, who has reviewed over 2000 (!) cases submitted by telepathology and who has reviewed the slides from SHCH, and Dr. Nina Hurwitz from Basel, who has reviewed all the slides for the NRH study and who has relentlessly tested many new features in our software.

I would like to acknowledge all those who have tested, translated and contributed otherwise to iPath and who provided me with invaluable feedback and new ideas. Thanks are expressed in particular to Dr. Gunter Haroske for continuously testing the clinical documentation module, to Fred Hersch for all the good ideas and discussions and to the "translators": Dr. Eka Kldiashvili (Russian and Georgian), Dr. Maria Zolfo (Italian), Dr. Oleksiy Solovyov (Ukrainian), Dr. Jorge Rodriguez (Spanish), Dr. Ugnius Mickeys (Lithuanian) and Dr. Bernard Tappernoux and Dr. Ousmane Ly (French). Most notably I would like to thank Monika Hubler, who has always kept our iPath-server running and organised while I was away and who has been a tremendous help in keeping our virtual institute going.

From all those many people who have contributed in various other ways I would like to speak a word of gratefulness to Afsin Abdi Rad, Mostafa Mohammad Golam, Christian Öhlschlegel, Cornelia Häner, Gernodt Jundt, Peter Dalquen, Udo Köllmann, Julio Claro, Geneviève Learmonth, Andrei Stepien, Boniafce Kabaso, Peter Scally, Markus Helfrich and to many others.

From all the friends that I have been discussing various aspects of this project I feel particularly indebted to Mark Wirdnam, Thomas Käppeli, Anthony Odama, Nhati Bangeni, Mohammad Khan. I would like to thank those people who have visited us during our year in South Africa: Kristin and Doris Roth, Christoph Bieri, Susanne Stolz and my parents-in-law, Sibylle and Jochen Haase. A very special thank must be addressed to Kenneth Chanda for the continuous discussion about telemedicine and ICTs in Zambia, his wife Catherine for the most incredible hospitality that we received when visiting them in Lusaka, and to Flora Asah for the numerous inspiring discussions on telemedicine and health information access in Africa. Furthermore I would like to thank all the wonderful people from Ikhwezi Lokusa, in particular Sr. Franziscus Maria and Sindiswa Mkwebetu and her family, who have made our stay in Mthatha a most enjoyable one.

At this place I also whish to express my profoundest thankfulness for all those software developers who helped to build all the open source tools without which none of this work would have been possible in the current form. The iPath telemedicine platform presented here is based entirely on open-source software as was the typesetting and statistical analysis of this thesis itself.

Finally, I thank the members of my family and most profoundly I would like to thank my dear wife Andrea, who has always supported my often chaotic activities and who has been the most caring and loving companion throughout the various journeys of the past years.

Financial support for the year visit at the University of Transkei was granted by the Swiss National Science Foundation. The telepathology project on Solomon Islands was made possible by financial support from the association "South Pacific Medical Projects" of Dr. Hermann Oberli and the Stanley Thomas Johnson Foundation.

Summary

In many developing countries there is an acute shortage of trained medical specialists. This does not only hamper individual patients' access to medical diagnostics but furthermore limits the development of health systems because a major role of the specialists is the provision of continuous medical education of health care personnel.

The rapid development of information and communication technologies has enabled radically new forms of virtual collaboration at a distance. So-called telemedicine enables us today to transmit knowledge to the patient rather than to only transport patients to the centres where the knowledge is available; this has promising implications in particular for remote and under-served areas.

Initiated by a request from a Swiss surgeon from Solomon Islands, a project for supporting the hospital in Honiara, capital of Solomon Islands, with pathology diagnoses was started between Honiara and the Department of Pathology in Basel in 2001. After a successful start this pilot project found broad interest, and the Internet platform that had been developed was soon utilised by projects from other countries and medical disciplines. Thus, questions arose about the diagnostic accuracy of such remote diagnoses as well as about their acceptance and impact on the local health care system. The work presented here was initiated on this background. It analyses the applicability of telemedicine in the context of resource-constrained areas and in particular the possibilities to extend its impact from improving individual diagnosis towards strengthening health care systems.

A central part of this project was the development of iPath, an Internet- and email-based telemedicine platform, which facilitates medical consultations, knowledge exchange and continuous education on a global scale. A particular emphasis was put on the applicability and accessibility for users from developing countries with limited infrastructure and network connectivity. The complete software was released under an open-source licence in order to allow unrestricted re-use for other institutions.

The diagnostic accuracy of this form of telemedicine was studied in two projects from the field of pathology. A retrospective review of over 200 glass slides from each project revealed complete diagnostic concordance between the telemedical diagnosis and review diagnosis in 69% and 85% respectively. Clinically relevant discrepancies were found in 8% and 3.3% of all examinations. Selection of images by the non-expert and communication were found to have the greatest impact on diagnostic accuracy. Both factors can be addressed by training and organisation of workflow. In comparison to submitting material for pathological examination by courier, the turn-around time could be reduced from weeks to days or hours. Besides the more rapid availability of diagnosis, telemedicine enabled a direct dialogue between the surgeon and the pathologist and thus facilitated an implicit permanent medical education.

The educational aspects of telemedicine were studied within the scope of a tele-dermatology project in South Africa. Distance collaboration with a dermatologist empowered a general practitioner based in a rural area to diagnose and treat a majority of patients with dermatological problems. Besides the direct benefit of saving the patients the cost of transportation to visit the dermatologist, the general practitioner could strengthen his own diagnostic skills under direct guidance and quality control of a specialist. As a consequence he will be able to treat more patients locally, close to their homes and families. The whole project was implemented within the local health system in order to facilitate a future inclusion of other primary care facilities.

Regional telemedicine networks play a major role to ensure relevance and acceptability of consultative and educational telemedicine. Within the scope of the Ukrainian Swiss Perinatal Health Program a telemedicine component was included, and it was found that the use of regional language as well as inclusion of the regional specialists are important for the acceptance of telemedicine and should not be neglected in a era of globalisation.

The presented results demonstrate that save and reliable telemedicine can be implemented with limited resources. Telemedicine is suitable in particular to strengthen existing international collaborations and to support professionally isolated medical specialists.

Regional collaboration and inclusion of regional specialists are desirable if telemedicine shall help to strengthen health care systems. The application of telemedicine should not only focus on providing care to individual patients, but should explicitly incorporate skills development and capacity building of primary care staff.

Organisation of work flow and communication have been found to be the most challenging task for the implementation of telemedicine networks. Resources must be invested not only in technology but more importantly in training and organisation. Utilisation of existing technological infrastructure is advisable wherever possible and greatly reduces the complexity of providing support and maintenance.

The presented telemedicine platform provides an efficient tool for the organisation of interdisciplinary, regional and international telemedicine networks. We hope that the unrestricted availability of the software developed during this project will enable other institutions to utilise it for their own purpose and that they will thus be able to allocate resources on the organisation of workflow rather than technology.

Part I.
Overview

1. Introduction

Many developing countries are facing various problems in delivering health care and medical services to their population – lack of funds and constraint resources as well as a dramatic shortage of trained and experienced doctors and nurses. Good quality services and medical specialists are often concentrated in urban areas. Poor roads, limited transportation facilities and long distances are severe obstacles for providing health care services to rural communities and remote areas. Patients can often not afford transportation to the nearest health care facility providing the necessary medical specialty. Provision of support and continuous medical education for those health care professionals working in rural areas are extremely difficult. Is telemedicine a feasible tool to address at least some of these issues?

In brief, telemedicine is the delivery of health care services such as diagnosis, treatment advice and continuous education for professionals at a distance, using Information and Communication Technologies (ICTs) to bridge the physical distance between patients and health care providers. In the industrialized part of the world, telemedicine is more and more regarded as a viable tool for delivering health care services in rural and remote areas. Considering the limitations of health care delivery in developing countries, the possibility to deliver medical services at a distance is an attractive idea and the concept of utilizing telemedicine in developing countries has been propagated for a long time[176]. Few pilot studies have shown that telemedicine may be applicable for the support of health care providers in developing countries. However, there are almost no examples of telemedicine networks that have been sustained beyond a pilot project and that have included more than just a few pilot sites. The only reports published in the scientific literature are from Malaysia[139] and India[5].

In the scope of this project we have developed an open-source telemedicine platform that simplifies distance collaboration between different health care providers whether they are in developing countries or in Switzerland. Usage of the platform was made available to many different projects – some within our own country but also to projects collaborating with various developing countries. This has allowed us to examine the feasibility of this form of collaboration and its implications for the involved partners in very different environments.

Why Telemedicine in Developing Countries?

We may ask who needs telemedicine in developing countries? Telemedicine is a form of delivery of medical services. Anyone needs access to medical services. However, the impact of "medicine at a distance" is not the same for everyone.

In countries where health care facilities are rare and where access to medical services is restricted by distance and poor transportation, telemedicine offers possibilities to distribute medical services more equally by utilizing ICTs. If travel costs for a patient to visit a medical specialist are higher than the cost of providing a telemedical consultation, then telemedicine might even be an economically affordable option.

Besides helping the patient, telemedicine offers possibilities for creating networks between the geographically separated health care providers. A functioning network is the basis for an efficient referral system and it is a good channel for disseminating information and education to improve local delivery of health care. Lack of communication between health care providers within a health care system has been identified as one of the prominent problems in many developing countries[12, 13, 78].

Additionally, health care delivery is often depending on humanitarian aid - generally in the form of funds and donated equipment but also of volunteers. Telemedicine enables another form of humanitarian aid whereby specialists from industrialised countries can offer part of their time and knowledge to assist doctors and other health care professionals in developing countries. For the development of medical services the collaboration with experienced professionals is often inevitable.

Why a Validation?

The United Nations World Summit on the Information Society 2003 listed e-Health as one of the key applications in its plan of action[175] and in 2005 the World Health Assembly debated on an e-Health policy[173]. Despite this enthusiasm, it is necessary to be careful when introducing new technologies. How easily can we abstract from the benefits of telemedicine and e-Health in Europe to its application in Africa? Are these technologies delivering valid and accurate results under circumstances that are quite different from where the technologies have been developed? There are big differences from a world where telemedicine is a welcome add-on to a world where the implementation of technology may be at the expense of providing access to safe drinking water.

Besides the accuracy and safety of technology there are questions about the organisation and the sustainability of telemedicine in developing countries. Telemedicine may enable a consultation between a patient in a remote village in Papua New Guinea and a specialist in the UK. But is this not undermining local health care system and the position of the regional specialists? When discussing the introduction of telemedicine in developing countries, its integration with the regional health care systems is an important issue.

This thesis includes case studies applying the same telemedicine technology in a different context and allows us to draw some conclusion about the validity and impact of telemedicine under these different settings.

Why Open-Source?

A major goal of this project was to create models for the application of telemedicine that can be reproduced easily. It is necessary to understand the prerequisites and conditions under which each model is applicable. However, the nicest model has only a limited validity if it cannot be reproduced easily. To facilitate and promote the reproduction of solutions emerging from this project, it was decided to release all software components developed during the project as free and open-source software. We hope that the free and unrestricted availability of the software used in our projects will allow interested institutions to create similar solutions more easily.

For many projects in resource-constrained areas the licensing costs for commercial products impose a high initial barrier. If unaffordable technology is donated to developing countries at discount rates, this bears the danger of increasing even more the dependency from donors. A possible alternative is often seen in Free and Open-Source Software (FOSS), which is made freely available by its developers – free as in "free speech" rather than in "free beer". The most important aspect of FOSS is the possibility to modify and adapt the software to regional needs. Local software engineers have full access to the source code of the open-source software and can modify it according to their needs. In addition, there is the possibility of integrating it with other health care technologies. As the market for health technologies in developing countries is often too small to be interesting for international companies, it is very unlikely that specific local requirements will be implemented in commercial products.

While creating a network for information and knowledge exchange between medical specialists is an explicit goal of this project, we hope that the unrestricted availability of the software will initiate a network of software developers and people interested in health care and telemedicine technology who will enable an exchange of technical information and that a continuous technology transfer will be another result of our activities.

Scope of this Work

This thesis is a combination of original research publications which were written during the PhD project and of chapters that glue these research papers into a concise framework. It is diveded into four parts: 1) overview of telemedicine and its current status, 2) development of a multi-purpose telemedicine platform with a special focus on pathology, 3) evaluation of several implementations and 4) a consolidation and discussion of the results with a focus on the health system context.

Part I gives an introdcution to telemedicine placing it within the general field of ICTs in medicine. Chpater 3 will highlight the most important definitions and conventions in the field and will give a brief overview of the specific context of telemedicine in resource-constrained areas. Chapter 4 will review the current literature focusing especially on practical telemedicine that involve some form of imaging and on its application in the context of developing countries.

Part II gives an overview of the iPath telemedicine platform, an open-source software that was developed during the PhD project. Although the development of this software is the central part of this PhD project, this thesis is not about computer science but rather about the public health aspects of telemedicine in the specific context of resource-constrained areas. An overview of architecture, design and functionality of the platform is given in chapter 5, while chapter 6 focuses specifically on the implemetation of the real-time telemicroscopy modul which serves as an illustration of the techncial details of iPath. The remaining chapters provide a descriptive overview of different applications of the iPath-server at the University of Basel – chapter 7 is a systematic overview while chapter 8 is focused on low resource settings.

The results of several of the telemedicine projects established within the scope of this PhD are presented in part III in form of published research papers or as working papers that are intended for subsequent publication in form of scientific articles.

In part IV these results are consolidated in a more general form in chapter 14. Based on these experiences I will then discuss telemedicine in the context of health systems in chapter 15. Before forumlating our conclusions I will insert a short chapter on ongoing projects which have been initiated on the background of the work presented here.

2. Goals and Objectives

2.1. Overall Goals

Development of an open-source software platform which simplifies the creation of telemedicine networks and the validation of the usefulness and feasibility of this form of telemedicine and its implications on health care delivery in the context of resource-constrained areas.

2.2. Objectives

1. Development of a multi-purpose open-source telemedicine platform for medical consultations, collaboration, knowledge sharing and educational applications.
2. Adaptation of the platform to needs and requirements of users in low resource settings.
3. Development of organisational framework for providing remote diagnostic consultations.
4. Creation of a global network of medical specialists willing to share knowledge and experience with colleagues in developing countries.
5. Evaluation of the user acceptance, diagnostic accuracy and educational value of this form of telemedicine.
6. Study the feasibility of this platform for building regional telemedicine networks within resource-constrained areas.
7. Determine potential impacts and benefits of telemedicine for a health care system.

3. Definitions and Context

Telemedicine – although referenced frequently – is not a clearly defined term. Basically it is medical practice at a distance using Information and Communication Technologies (ICTs) to overcome the distance between the partners involved. However, before attempting to define the term more rigorously it is useful to place telemedicine within the general application of ICTs in health care as there are many overlapping areas and possible interactions.

3.1. Information and Communication Technologies in Medicine

Over the past two decades an immense proliferation of Information and Communication Technologies, commonly abbreviated ICTs, could be observed. Communication networks from plain old telephone lines to mobile phone networks and satellite communication have reached almost every corner on our planet. The Internet has become a global repository of information and almost any type content from daily news papers over share prices at the stock exchange market to specialised scientific journals are all accessible over the Internet. And finally, today's desktop computers are capable of handling complex multimedia content such as images and also movies with ease, allowing the production of digital content for basically everyone who can afford a computer. ICTs have penetrated almost all aspects of our lives.

In the health care sector there are many different applications of ICTs. The most common is probably the management of administrative data such as billing and general record keeping – areas in which probably most of us do not even remember the time before ICTs. Besides, the electronic management of patient information is becoming more and more important and many hospitals are working towards digital storage of all patient associated data using electronic medical records (EMR) or electronic health records (EHR). But also in medical practice itself there are many new methods that directly depend on ICTs. In modern medical imaging such as CT (Computer Tomography) or MRI (Megnetic Ressonace Imaging) but also in standard radiology (e.g. plain thorax x-ray) the conventional, film based equipment is more and more replaced by digital radiology (DR), film-less solutions, in which all image data are primarily stored in electronic form and only transferred to film for reading in locations not (yet) equipped with digital X-ray viewing stations.

Another field where ICTs have become unavoidable is in the very heart of modern evidence based medicine – today, access the evidence base has become almost impossible without the help of ICTs. Scientific articles are searched through pubmed, the on-line database of the National Library of Medicine. Articles are then accessed through "virtual libraries" in form of PDF (Portable Document Format) documents. Medical evidence databases such as e.g. Cochrane are commonly accessed over the web. Besides, there is an overwhelming amount of data accessible through the world wide web, ranging from electronic teaching aids to on-line patient forums for almost any kind of disease.

3.2. Telemedicine

3.2.1. Definitions

Telemedicine is not a concisely defined term. Some authors use it in the stringent sense of the word derived form the syllable "tele" applying telemedicine strictly to the practise of medicine "at a distance", which often involves a consultation between a patient and a geographically separated doctor, using a video conference link. Many others however use the term telemedicine in a much broader sense of "rapid access to shared and remote medical expertise by means of telecommunications and information technologies, no matter where the patient or relevant information is located" (EU Commission on health care telematics).

Besides the term "Telemedicine" many other similar term are frequently encountered in the literature – e.g. "Health Telematics", "Tele-health", "Tele-Care", "On-line Health", "e-Health", "Medical Informatics" or simply "ICTs for health". The terms for describing the same phenomenon are extensive and the terms are not used in any precisely defined way.

The Norwegian Centre for Telemedicine (NST), WHO collaboration centre for Telemedicine, is using the following definition: "Telemedicine is the investigation, monitoring and *management* of patients and the *education* of patients and staff using systems which allow ready access to expert advice and patient information no matter where the patient or relevant information is located".

Telemedicine encompasses a broad range of applications and services. Each component involves different providers and consumers. To facilitate the understanding of telemedicine it is useful to distinguish between the service to be delivered and the delivery mechanisms. The following overview is derived and extended from the definition of telemedicine of the American Telemedicine Association (ATA)[1].

Services

- **Specialist Referral Services** typically involves a specialist assisting a general practitioner in rendering a diagnosis. This may involve a patient "seeing" a specialist over a live, remote consultation or the transmission of diagnostic images and/or video along with patient data to a specialist for viewing later. Routine applications of specialist referrals include all medical disciplines that are to some extend based on visual data such as radiology, pathology, dermatology, ophthalmology, cardiology, etc.

- **Primary Patient Consultations** between a patient and a primary care or specialty physician with the aim of rendering a diagnosis and treatment plan using transmission of audio, video and medical data. Such a consultation might be in the from a telephonic consultation or may include communicating to a physician over the Web.

- **Remote Patient Monitoring** uses (miniature) devices to remotely collect and send data to a monitoring station for interpretation. Such "**home tele-health**" applications might include a specific vital sign, such as blood glucose or heart ECG or a variety of indicators for home-bound patients. Such services can also be used to support visiting nurses.

- **Medical Education** provides continuing medical education credits for health professionals and special medical education seminars for targeted groups in remote locations.

- **Patient support service** include reminders to take medication, supervision, scheduling of appointments and similar applications which are not implicitly medical but which are important to improve the outcome of care.

- **Consumer Health Information** includes the use of ICTs for consumers to obtain specialized health information and on-line discussion groups to provide peer-to-peer support. This is often regarded as part of "**e-Health**"

Delivery Mechanisms

- **Networked Programs** link tertiary care hospitals and clinics with outlying clinics and community health centers in rural or suburban areas within an existing or newly created network of health providers. The links may use dedicated high-speed lines or the Internet for telecommunication between sites.

- **Point-to-point Connections** using private networks are used by hospitals and clinics that deliver services directly or contract out specialty services to independent medical service providers at ambulatory care sites. Radiology, mental health, pathology and even intensive care services may be provided under contract using telemedicine to delivery the services.

- **Primary or specialty care to the home connections** involve connecting primary care providers, specialists and home health nurses with patients over single line phone-video systems for interactive clinical consultations.

- **Home to monitoring center** links are used for cardiac, pulmonary or fetal monitoring, home care and related services that provide care to patients in the home. Often normal phone lines are used to communicate directly between the patient and the center although some systems use the Internet.

- **Web-based** systems are becoming increasingly popular for two main applications. 1) e-Health patient service sites provide direct consumer outreach and services over the Internet. 2) Communities of specialists who share medical data over the web for second opinions consultations and for continuous medical education.

- **Messaging-based** system are useful for sending out reminders or quick delivery of laboratory results to smaller health care facility. Messaging services such as SMS or email do often not have a guaranteed delivery and are thus not advisable for critical information, but the are easy and cheap to deploy on a large scale.

Involved Partners

In extension to the original list on the ATA website it is useful to also consider the partners involved in the telemedical applications.

- **Specialists** who do second opinion consultations, sub-specialty consultations, consensus diagnosis or joint research.

- **General Health Professional and Specialist** – referral to establish diagnosis and also educational applications and decision support (e.g. tumour board meetings)

- **Patient and Health Care Professional** are involved in primary consultations, e.g. pre-clinical consultation over telephone (medical call centres, in Switzerland for example services such as Medgate[1]) as well as for targeted information delivery to patient such as reminders to take medication.

- **Patients** use ICTs for exchanging information and knowledge e.g. in on-line forums. This is increasingly used by "self help" groups on issues about living with a certain disease or disability.

- **Specialist to Computer Applications** are used for remote quality assurance, decision support and joint research.

- **Patient to Computer Applications** such as health awareness web site are used to disseminate information to patients and to provide preventive information.

3.2.2. Technology

Telemedicine is not necessarily a new and complex technology. The first telemedical consultation probably dates back to the time when the telephone was invented. Even today, consultations over telephone are probably the most frequent form of tele-consultations between patient and health providers.

---

[1] Medgate is a private medical call centre operating in Basel, Switzerland. http://www.medgate.ch

However, with the advance of ICTs new forms of data transmission have become available. One technology that has certainly impacted telemedicine is video conferencing. The life transmission of image and voice of two partners has often been regarded as a key necessity for telemedicine consultations. In reality, the importance of video conferencing is depending on the specific type of telemedicine. While video conferencing is ideal for a tele-psychiatry consultation, video conferencing technology have never been widely used e.g. in Pathology. One often overlooked limitation of video conferencing is that the consultation must necessarily be held in **real-time** with all partners sitting together at the same time. With busy schedules of modern medical work the planning of appointments with all partners can be difficult, especially if some partners are not located in the same timezone. Additionally, the life use of technology makes the application also prone to technical problems. Technical problems at one partners side tend to block everyone else.

An alternative to real-time consultations is so called **store-and-forward** telemedicine where the partners work in an asynchronous way. The non-expert or submitter will prepare his question and the necessary material (store) and then transfer it to the expert for review (forward). The expert can review the material at a later time and write a report back to the non-expert. A consultation sent by email is a very basic form of store-and-forward telemedicine.

With the medical knowledge constantly increasing and diagnostic methodology as well as treatment options multiplying, clinical decision making is getting more and more complex. Involvement of appropriate sub-specialists and communication between different partners in the health care systems are becoming more and more important. Email is often not appropriate to organise structured communication in a larger group. Thus, specialised often web-based applications are more readily being used for telemedicine. In contrast to email and especially to video conferencing, web-base and database driven applications simplify the archiving of the telemedical process.

Some specialties that have more complex needs on data transmission have developed standards how to transmit medical data in digital form. In radiology the DICOM (Digital Imaging and Communication in Medicine) standard defines how to store a digital X-ray and how to transmit it form one health provider to another (c.f. section 4.3.1).

3.3. Specific Aspects for Resource-Constrained Areas

Modern ICTs and worldwide communication through the Internet promise universal access to information and the globalisation of the medico-social network's modes of communication between doctors, laboratories, patients, and other health professionals. The term "globalisation" however has to be used with care as it often ignores the reality of the "digital divide", that is, the fact that social inequalities may preclude the realization of this promise on a truly global scale[121]. In many aspects globalisation can even be seen as a force creating various kinds of devide, among them, economic devides as well as digital devides. Globalisation is often resultaing Furthermore, without local ownership and adaptation, application of ICTs exhibits a strong tendency to normalise cultures towards our western culture.

Conceptual work on telemedicine and the involved technical tools are almost exclusively developed the industrialized nations. For the application of telemedicine in developing countries, where resource are often extremely limited, it is thus imperative consider the socio-economical differences of these counties and their health care systems compared with the industrialized part of the world.

3.3.1. Potential Benefits

Many developing countries have an acute shortage of doctors, particularly medical specialists such as dermatologists, radiologists or pathologists. Sub-Saharan Africa has, on average, fewer than 10 doctors per 100 000 people[50]. Moreover, the distribution of health professional is often very unequal and specialist services are mainly concentrated in a few urban areas. Doctors and nurses in rural health centres, who serve most of the population, lack access to adequate health information, the lifeblood for the delivery of evidence based health care services[3, 109]. They are isolated from specialist support and up to date information by long distances and poor roads and poorly established use of telecommunication.

Information and Communication Technologies (ICTs) do have the potential to help over come many of these limitations. Telemedicine may help rural health workers and their patients in different ways. For many common health problems it is possible to establish a diagnosis and a feasible treatment plan using telemedicine[22, 130]. Tele-consultations can play an important role in triage decisions, establishing the need for the physical referral of a patient. Transport to the referral centre is often at the expense of the patient and many patients in developing countries cannot afford (unnecessary) travel to a next level hospital[108]. Telemedicine offers opportunities for continuous medical education in remote areas which will benefit in improved local health care delivery[53]. ICTs have a great potential of enabling access to relevant and up-to-date health information[2] and e-Learning[53] and finally, ICTs may also help to improve the administrative process in clinics and thus free human resources to be spent with patients[129].

3.3.2. Limitations

Computers and Connectivity

In the whole sub-saharan region of the African continent there are probably less Internet users than in New York City[121]. It is estimated that the whole African continent has about 24 million Internet users which is equivalent to 2.6% of the population[100]. North America alone has 230 million Internet users (69% of the population). The availability of electricity and especially of communication lines are very limited which in turn restricts the number of locations in which telemedicine could be easily deployed. Besides, compared to the per capita income computers are very expensive and are generally not available to individual users. On average there are 32 telephone lines, 60 cellular subscription an 17 Internet users per 1'000 people[79].

One effect of the low usage of computers and Internet technologies are the relatively high prices for connectivity. In South Africa the costs of Internet connectivity is 10-100x higher than in most European countries. Instead of making Internet services cheap and available as a mass product, service providers are often targeting the few wealthy that can afford a computer and connectivity – e.g. foreign companies, embassies, NGOs (Non Governmental Organisations) – at prices beyond the reach of ordinary citizens.

Another problem is the lack of decent bandwidth. Most Internet connections are very slow and chronically unreliable. For example, the Walter Sisulu University in Mthatha, South Africa has an Internet connectivity if 512 KB/s serving the whole the whole campus. This is less than a standard private ADSL (Asynchronous Digital Suscriber Line) connection in Switzerland (600-2000KB/s). And South Africa is certainly not the least developed country in Africa. A further limitation for the deployment of real-time telemedicine applications is that most affordable communication links are of an assymetric nature: with many satellite connections and in particular with ADSL the upload chanel is often too slow to allow high quality real-time communication.

Not to be neglected is the fact that in telemedicine, the flow of information is usually bidirectional. A satellite connection that allows decent speed in receiving of information may be extremely slow for transmission of data. Technology that rely on stable and fast connections are very much bound to fail. By far the most feasible technology in many places is simple email using the SMTP (Simple Mail Transfer Protocol) which is very robust and tolerant to unreliable network connections as it was developed in the 1970s, a time when (temporary) dial-up connections were still commonly used to connect even larger institutions with each other.

Computer Literacy and Support

Although most telemedicine concepts discussed in this thesis are not targeting directly the individual users but are rather allocated on an institutional level, the lack of computer means that most of the key personnel – doctors, nurses, health officers – are not very familiar with computers. More importantly, a lack of computer users also prevents the development of (commercial) support services. In many developing countries it is extremely difficult to find trained staff for the installation and maintenance of computers and computer applications and thus those using computers must often rely on themselves to fix problems.

Information Readiness

The limitations are not only of technical nature. At a ratio of less than 10 doctors per 100 000 people[50], these few doctors have only very limited time available besides consulting patients. Many doctors working for public hospitals are not able to live decently from the governmental salaries and are running private chambers.

The use of scientific literature and evidence to improve health care is regarded as common practise in modern, evidence based medicine. It should however not be overlooked that access to scientific literature alone is often not sufficient. Especially for those health professionals who are not fluent in English the use of international scientific literature poses a high initial burden. While students in most industrialized countries are taught the importance and the use of literature at University levels this is normally not the case in many developing countries. Most people are not naturally "information ready", but need some guidance and tutoring to learn how to deal with the vast amount of of information available today, how to "translate" evidence and information into daily practise and how to select between trustworthy and doubtful sources of information.

4. Literature Overview

This chapter tries to give a brief overview of telemdicine in the current research literature. Starting from a very short review of the evolution of telemedicine this overview will primarily focus on the situation in developing countries taking Wootton's article form 2001[168] as a starting point.

Based on similar searches on medline an overview of different fields of telemedicine without limitation to developing countries is given; the results are summarised in table 4.1. For the sake of siplicity I will concentrate on practical application of telemedicine in fields that are relevant within the scope of this thesis.

4.1. Evolution of Telemedicine

While interest in telemedicine has increased in the last few years due to recent advances of ICTs, the concept is not new. The first reference of the subject is probably the famous "Radio Doctor" cover image of the 1924 Radion News Magazine[54]. One of the first telemedicine applications reported in the scientific literature was probably the project for transmission of radiologic images by telephone between West Chester and Philadelphia, Pennsylvania, a distance of 24 miles[55, 56]. In the 1970s the number of telemedicine projects started to grow and first real-time applciations are mentioned[43, 103]. The STARPAHC Project, for example, tried to introduce telemedicine in a the rural Papago Indian Reservation in Arizone[51]. Throughout the 1980s telemedicine speciality specific applications started to emerge, for example telepathology, which was first mentioned in 1986[151]. In the field of radiology saw the development of the first standard on digital medical imaging which culminated in the release of the DICOM specifications in 1992[9]. The number of telemedicine applications started to grow rapidly in the 1990s due to availability of Internet and affordable computers and digital imaging solution and the latest technical breakthrough of telemedicine was probably the first transatlantic robotic operation which was performed in 2001 by a surgeon in New York on a patient in Strasbourg[93].

4.2. Telemedicine in Developing Countries

Five years ago, when our project was just starting with the very first collaboration with a developing country, Richard Wootton wrote in an article on "Telemedicine and Developing Countries"[168] that "despite continued interest in the use of telemedicine in the industrialized world, it cannot yet be considered to have entered in the mainstream of health-care". Has that changed in the past five years since the publication of Wootton's article? And has the situation changed in the developing world?

In Wootton's article, a literature search on Medline[1] for the term "(telemedicine OR telehealth OR teleradiology) AND (developing world OR developing country)" yielded 39 articles. Today the same query will find 129 articles or even 145 when the plural for developing countries is added. From these 145 published articles 17 are labelled as review articles by Medline. From numbers alone this looks like a significant increase in activity in this field. If we look at the more recent literature – 91 articles found in Medline that have been published since 2001 – we see that many of these publications are either comments or purely argumentative articles. 11 articles are reviews and 19 articles are commentaries. 19 articles are descriptive reports of existing projects and programmes, four are reports about pilot projects and three are technical descriptions, two are about needs assessments, two are dealing with scientific publishing and one with telehealth policies. 13 papers are evaluating activities and outcome in telemedicine projects in developing countries. Out of these 13 papers, four are evaluations of the Swinfen Telemedicine programme, two are evaluating the Partners Health telemedicine project in Cambodia. A further two publications are evaluating the accuracy of diagnostic X-ray imaging, two articles evaluate tele-dermatology projects in Mexico and Pakistan, one is an evaluation of the accuracy of tele-cytology, one is an evaluation of a project using VHF radio in rural areas of the amazon region and one is a study on quality of telepathology diagnosis comparing two European telepathology services with cases submitted from Iran.

There are, however, other publications that are not retrieved by a query limiting to the term "developing countries". Table 4.1 illustrates that searching for telemedicine and a certain region such as Asia, Africa or India reveals some additional publications. Further there are other specialties besides teleradiology such as pathology or teledermatology. But even considering those gives only a weak scientific evidence on the validity and efficiency of the application of telemedicine in developing countries. Some further evaluations of the medical outcome and the validity of telemedical diagnosis are presented in part III of this thesis and will also be published as scientific articles.

It is also important to bear in mind that there are many more telemedicine activities in developing countries. The report of the ITU study group on telemedicine in developing countries[70, 177] delivers over 150 pages with descriptions of telemedicine programmes from 22 countries. A look at the programmes of large international conferences such as Med-e-tel in Luxembourg or Medinfo (this year in Toronto) clearly demonstrates that there are a lot of activities, which are not published in scientific journals. Finally, a query on Google about "telemedicine AND developing countries" retrieves approximately 230'000 Internet pages!

---

[1] http://www.pubmed.gov/

4.2.1. Impact of Telemedicine in Developing Countries

A fundamental issue with telemedicine in developing countries is to study the impact of telemedicine on the health care system and on society in general. In a letter to the British Medical Journal, Rigby [127] asked: "Will strengthening secondary care for a few disadvantage basic primary care or environmental health for the many? Will investment in the required rural telecommunications be at the expense of providing drinkable water? Will developing countries too be seduced by the expensive impact of technology led tertiary care for the few, while ignoring the endemic impact of modified health related behaviour?"

These aspects are very fundamental in any kind of international collaboration, especially in the field of ICTs. Most ICTs-based projects will in fact deliver information and knowledge, not safe drinking water itself. But the information and knowledge delivered may turn into a driving force for people to establish safe drinking water as the knowledge about communicable disease will teach them why safe drinking water is so important. While telemedicine may indeed bring *direct* benefits only for the few in secondary care, it also has the potential to strengthen the local health provider in improving the services offered and thus benefit the general public in a more indirect way[2, 130, 179, 181]. It is thus important to study the impact of telemedicine on health care systems and societies in developing countries. We should not forget that knowledge and experiences on how telemedicine, e-health and ICTs in general may change health care and access to medical information and continuous medical education in the industrialized world is not necessarily reflecting the situation in countries with limited resources[121].

In the scientific literature, reports on the impacts of the application of telemedicine in developing countries are almost entirely absent. Brandling-Bennet et al.[15] report on a project in rural Cambodia. During the first 28 months 264 consultations were examined. The major impact reported is a reduction of the "mean duration of chief complaint at initial visit" from 37 months before the introduction of telemedicine to 8 months at the end of the study. Additionally, the percentage of patients referred for care outside of the village could be reduced from almost 80% to under 20%.

One impact that is cited by different authors are the cost savings for the patient due to reduced travelling[5, 147]. Another impact is the possibility to use telemedicine to transfer knowledge that may finally result in more cost effective treatment of patients and thus help to save scarce resources[22, 75, 182].

Interestingly, the majority of publications on telemedicine in developing countries are in fact focusing on the application of telemedicine for collaboration between doctors and nurses in low resource settings and their partners in industrialised nations. A notable exception is the work of Geissbuhler et al.[53] who evaluated the feasibility, potential and risks of an Internet-based telemedicine network in developing countries of Western Africa. In an evaluation of the pilot project between Mali and Geneva they come to they stress the importance of fostering of South-South collaboration channels and the valorization of local knowledge and its publication on-line. A second phase of their project is now focusing on creation of decentralised networks within l'Afrique Francophone.

4.3. Medical Specialties

Telemedicine is not a defined technology of its own, but the utilisation of ICTs to separate a medical process over geographically distant partners. The problems involved with this separation are different from one medical specialty to another. To understand these implications, not only for developing countries, it is most useful to look at the general literature on telemedicine. A literature search on Medline[2] reveals 7845 entries for the keyword "telemedicine" – using the MESH term "telemedicine" retrieves 7435; roughly the same amount. 806 of these citations are review articles and 457 also include the key word "clinical trial". In order to refine this amount of literature, I have made a number of queries for different medical disciplines such as radiology, pathology or dermatology using e.g. the keyword "telepathology AND tele-pathology". The results of this search are summarised in table 4.1.

While Medline provides very useful Medical Subject Headings (MeSH) for many fields of medicine, in the field of telemedicine only the terms "telemedicine", "teleradiology" and "telepathology" are defined in the MeSH database. Thus, most of the following will be based on plain keyword queries. One way of using the MeSH terms is combining the MeSH term for the medical speciality and the keyword telemedicine. Telepathology translates to the search term "telemedicine AND pathology[MeSH]". Table 4.1 lists the results of these queries. Rows 1-5 reflect the results from the keyword search while the results using MeSH are listed in last column. For most categories there is no substantial difference although using MeSH terms is generally a bit more restrictive than keyword search. The only notable exception is tele-psychiatry where the the search on "telemedicine AND psychiatry[MeSH]" retrieves over twice as many articles as a keyword search on "telepsychiatry OR tele psychiatry". This may indicate that the term telepsychiatry has not yet been established as a naming for the application of telemedicine in psychiatry.

In the following, I will focus on those specialties that are based on some form of imaging – radiology, pathology and dermatology – as these are the fields in which I could gather some experience with the telemedicine platform developed in this project.

4.3.1. Radiology

The most prominent telemedical application is certainly teleradiology with over 1100 references in the scientific literature indexed on Medline (838 using the MeSH term "teleradiology"). This is not so surprising because in radiology, the basis of the diagnostic process is an artificially created image itself. Modern X-ray generate this image in digital form and with DICOM[9, 10, 46, 86, 110] there is an agreed standard format for these digital radiographs. Since such images are captured digitally, it is a logical step forward to also transmit them over a network and DICOM – standing for "Digital Imaging and Communication in Medicine" – includes a component for standardised transmission of images. In many modern hospitals this technology is in fact used to transmit the image from the actual X-ray machine to the screen in the radiologist's office. The transmission over a longer distance is only a logical consequence.

---
[2] 5 April 2006

Table 4.1.: Literature search on Medline from 5.4.2006. The column "medline hits" indicates how many entries are found in Medline using the keyword search given as "search term". The Column "[MeSH]" gives the number of entries when searching Medline for a combination of telemedicine and the speciality as MeSH term ("telemedicine AND specialty[MeSH]").

topic	search term	medline hits	reviews	%	clinical trial	%	[MeSH]
speciality	telemedicine	7845	806	10.3%	457	5.8%	7523
	telepathology OR tele pathology	619	87	14.1%	33	5.3%	470
	telecytology OR tele cytology	134	10	7.5%	11	8.2%	5
	telehematology OR tele hematology	4	1	25.0%	0	0.0%	10
	teledermatology OR tele dermatology	177	21	11.9%	32	18.1%	172
	teleradiology OR tele radiology	1104	119	10.8%	32	2.9%	400
	teleophthalmology OR tele ophthalmology	65	5	7.7%	3	4.6%	93
	telesurgery OR tele surgery	293	50	17.1%	22	7.5%	550
	telemonitoring OR tele monitoring	284	24	8.5%	36	12.7%	203
	telecardiology OR tele cardiology	87	12	13.8%	6	6.9%	78
	telepsychiatry OR tele psychiatry	157	22	14.0%	9	5.7%	356
	telehealth	555	60	10.8%	23	4.1%	
	telemedicine AND trauma	280	22	7.9%	23	8.2%	238
	telemedicine AND emergency	623	56	9.0%	43	6.9%	533
	Telenursing OR tele nursing	84	11	13.1%	5	6.0%	272
	telemedicine AND (HIV OR AIDS)	79	5	6.3%	6	7.6%	33
	teleteaching OR tele-teaching OR e-learning	178	16	9.0%	3	1.7%	
learing							
	telemedicine AND continuous medical education	9					
	sum		4732				
regions	telemedicine AND developing countries	114	16		1		
	(telemedicine OR telehealth OR teleradiology) AND (developing world OR developing country)	145	17		3		
	telemedicine AND africa	56	6				
	telemedicine AND asia	313	26				
	telemedicine AND india	48	6				
	telemedicine AND china	41	3				
	telemedicine AND russia	52	2				
	telemedicine AND cambodia	4					
	telemedicine AND solomon islands	13					
	telemedicine AND ukraine	5					
	south africa	25	2				

A second important consequence of the fact that the image itself is the basis for the diagnosis consists inof a very straightforward process of telemedicine in radiology. A transmission of the image in sufficient quality and time alone is enough for a tele-diagnosis and there is often no need to transmit additional data or perform additional examinations immediately. Thus, teleradiology lends itself to a separation of the patient and the specialist as all the specialist needs to see is the image of the patient. It is not surprising that teleradiology is the only application in telemedicine that is used on a larger scale. For example, the Arizona Telemedicine Programme has reported over 85'000 teleradiology events (88% of all telemedicine events) since its inception[89]. There are even companies in India offering overnight radiology reading for US hospitals, making use of the difference in time zones[118].

While digital X-ray equipment is increasingly used in industrialized countries, the vast majority of X-ray machines in the developing world is conventional film-based equipment. Teleradiology based on film based images is also possible and involves either scanning of the film or digitization with a digital photo camera. While scanners are relatively expensive especially for smaller hospitals in developing countries, capturing images with a digital camera is a very cost effective option. There are a few studies which evaluated the sensibility and specificity of tele-diagnosis on digital snapshots of standard radiology imaging (chest, abdomen, and bones)[26, 143]. This form of teleradiology is also frequently used by many projects on the iPath server at the University of Basel. Many of the consultations within the Ukrainian Swiss Perinatal Health Programme include digital photographs of X-ray and ultra sound images; also some of the pathology consultations are accompanied by digital photos of radiographs (e.g. bone tumour working group).

4.3.2. Pathology

After teleradiology, telepathology is probably the most frequently cited application of telemedicine (619 citations) and is the only other application that is listed in the MeSH index (427 entries in Medline). While in absolute terms the figures are comparable, the frequency of citations for telepathology is much lower if we compare the total amount of literature on pathology compared to the amount on telepathology. While in radiology, articles on tele-radiology make up 3.3% (1104 articles) of all publications on radiology (33'430 articles), this ratio is almost ten times lower in pathology. Out of the 145'620 articles on pathology, only 619 or 0.4% are dealing with telepathology.

The transmission of a medical problem in pathology is much more complex than in radiology. The diagnostic basis in histology is a thinly cut piece of tissue ($\sim 2-10 \mu m$) which is stained embedded on a glass slide and then viewed under the microscope at a magnification of up to 400 times. The amount of information contained in such a slide is huge and cannot be easily transmitted in digital form from one place to another. There are basically three ways of performing a remote diagnosis on a pathology slide.

4.3.2.1. Telepathology Systems

Static-Image Telepathology

In static image telepathology, a small set of 5-20 characteristic images are captured by the non-expert and then transmitted in electronic form such as email or over a web server and are reviewed by a distant specialist. This method is easy to implement but has its severe shortcomings. The specialist must rely on the proper selection of images by the non-expert. If additional images are required, this has to be communicated back to the non-expert and this process will take time. Static-image telepathology is, however, still the most frequently used method and under certain circumstances will deliver quite good results [39, 123, 150, 163, 178].

Dynamic Robotic Telepathology

Dynamic robotic telepathology or real-time tele-microscopy was introduced in the late 1980s in Arizona by Weinstein et. al [151, 155, 156, 152] and soon found early applications in Norway[45, 112] and also in Switzerland[114, 115]. Dynamic robotic telepathology systems consist of a motorized microscope which can be operated from a distant workstation. The image from the microscope is captured with a video camera and is transmitted in (near) real-time. The specialist can view the video image on his workstation and can remote control the microscopy, especially the location of the slide and the magnification. While dynamic robotic telepathology is giving full control of the microscope to the pathologist, it requires sophisticated and delicate technology and is relatively expensive.

All the early systems were using dedicated communication lines for the transmission of data: ISDN, ATM or satellite connections. In addition, there is no standard communication protocol and systems from different vendors were not compatible. These severe limitations prevented the large-scale application of telepathology in routine services, outside of a few research projects for a long time[32].

As an alternative to point-to-point tele-microscopy, systems using Internet based connectivity were developed by several groups[16, 62, 122, 144, 165]. The use of basic Internet technology (HTML/Javascript, Java or Macromedia Flash) basically turns an ordinary web browser into the tele-microscopy remote control. This eliminates the necessity to install a special software and dedicated communications lines in order to remote control the microscope. Additionally, there is the potential of sharing a remote microscope among many distant users. Some newer telemicroscopy solutions are now using this approach as standard way of remote control (e.g. Nikon CoolScope microscopes[3]). A major limitation with tele-microscope remains the fact that the original slide must be placed under a real microscope before the distant specialist can review the slide.

The tele-microscopy solution developed at University of Basel within the frame of this projects is discussed in detail in a separate chapter below.

---
[3]http://www.coolscope.com/

Virtual Slides

"Virtual slides"[34, 36, 57, 58, 157] are the most significant recent development in telepathology. This technology comprises the digitization of a whole slide which can then be stored and transmitted in digital form. This is achieved by scanning the whole slide at high magnification and then combining all images into a large mosaic. In order to record all information in a microscopic slide, it is often necessary to capture images at different focus levels.

The main limitation of virtual slides is the large storage capacities necessary and associated with this the problems of transmitting such large amount of data over a network. The digitized representation of a slide typically requires between 0.5-5GB of storage. However, with the ever increasing storage capacities of modern computer technologies and improved scanning methods that allow scanning of a slide in a few minutes, virtual slides have become manageable[157].

For teaching purposes "virtual slides" are quickly becoming the standard way of distributing microscopic slides for students[57, 66, 85, 154]. Many universities are no longer replacing their sets of teaching microscopes but have started to digitize their collection of teaching slides and making them available to the students over the Internet. In diagnostic pathology, there are no reports yet about the application of virtual slides. However, it is anticipated that this will change over the coming years[57].

4.3.2.2. Cytology and Hematology

Most reports on telepathology are certainly from the field of histopathology. Cytopathology and also hematology and hematopathology are other fields that could potentially benefit from telemedicine. The few available results on the application of tele-cytology[24, 83, 125, 178] and tele-hematology[90] show very promising results. Due to the simple laboratory technology involved, tele-cytology may be a very interesting application for developing countries. On Solomon Islands e.g. there is a high prevalence for cervical carcinoma[94, 104, 120]. An early diagnosis and thus regular screening of patients is required for a successful treatment[76]. Tele-cytology could help to develop skills for screening cytological slides and it would be particularly useful for quality assurance.

4.3.2.3. Telepathology in Developing Countries

For developing countries there are very few reports about telepathology. A search on Medline reveals only descriptive articles about the application of telepathology in developing countries[18, 38, 101, 135, 149]. In addition, there is a review about telepathology in India [6] and one article describing a study design simulating the diagnostic support from remote access to health workers in developing countries[27]; an article from Ethiopia gives a clinical case report of a telepathology consultation[132].

Telepathology in developing countries will be a major topic throughout this thesis and there will be several chapters on this issue below.

4.3.3. Dermatology

Dermatology is a medical discipline which is not based on a special imaging technology like radiology or pathology. However, the diagnostic process is very much based on the visual appearance of the skin and thus the process of tele-dermatology is relatively straightforward. The literature indexed on Medline contains 177 articles on teledermatology: 21 reviews and 32 clinical trials. Articles on clinical trials make up 18% of all articles and thus teledermatology is certainly the most vigorously evaluated discipline of telemedicine.

Unlike telepathology, teledermatology is normally practised in direct collaboration with a patient. Often, teledermatology is used to allow a general practitioner (GP) to consult with a dermatologist before physically referring a patient[4, 22, 68, 81]. Consequently, many of the clinical trials are looking at the patient satisfaction and acceptance of teledermatology by GP[30, 31, 37]. Clinical accuracy of teledermatology diagnosis versus face-to-face visit were reported between 48% and 90% [4, 91]. Reports on satisfaction with patients and GPs are not very conclusive. While some studies suggest high satisfaction[31, 161], a study by Collins et al. reports satisfaction only for 21% of 42 surveyed GPs [30].

With the availability of relatively cheap but high quality digital photo cameras, tele-dermatology seems to be pre-destined for application in low resource settings. Especially with the HIV epidemic, complex dermatoses are increasingly prominent in many countries. However, there is not much literature about the application of teledermatology specifically in developing countries. There are reports on diagnostic accuracy of teledermatology in Pakistan[126] and Turkey[119] and some reports on acceptance and impact of teledermatology for GPs in Africa[22, 28, 131]. A very noteworthy application of telemedicine is presented by Caumes et al.[28], where teledermatology was used in Burkina Faso to diagnose dermatoses in travellers (80% westerners). Besides the immediate benefit for the patients, the authors report that teledermatology has an educational value for local GPs.

4.3.4. Other Disciplines

Besides the areas covered in this thesis, there is a multitude of other applications of telemedicine. Very frequently cited in the scientific literature is the application of telemedicine in **emergency care** – "telemedicine AND emergency" retrieves 623 citations. In emergency situations the relatively quick availability of relevant knowledge how to treat the patient is of crucial importance. As time and locations of emergency cannot be predetermined, telemedicine is often seen as a means to bridge the gap between patient or mobile emergency teams and the specialists located at the hospitals [77, 146, 166]. The majority of articles on telemedicine in emergency care are certainly about stroke patients and are technically often related to the transmission of ECG data (**telecardiology**). Besides, telemedicine is considered for assistance on long distance transport such as flights[48] or ferries[74] and a very popular topic seems to be the delivery of telemedicine on space flights (79 citations).

Other applications cited in the literature are **telesurgery** (293 citations) and **tele-monitoring** (284 citations) of home care patients. A very valuable application of telemedicine for planning surgical interventions in developing countries was reported be Lee et al[84]. With the advancing miniaturization of electronic sensors and communication devices (mobile phones), home monitoring of chronically ill patients is quickly becoming a reality. **Telepsychiatry** is also relatively frequently cited (157 citations). Monnier et. al [102] give a good overview of the current state of telemedicine in psychiatry. Most publications identified were either clinical demonstrations, program descriptions and patient satisfaction studies. They conclude that "... Despite the rapid increase in information on telepsychiatry, methodologically sound studies in the area of telepsychiatry are still infrequent.".

4.3.5. Telehealth or e-Health

Besides these telemedicine applications in the narrow sense of delivering diagnosis or treatment advice for an individual patient, there is a much broader field of applications of ICTs in health generally summarised under the term of e-health or telehealth. Reports on e-health applications that are of particular interest for developing countries include examples such as "Compliance Service uses SMS technology for TB treatment in Cape Town region"[23] or the delivery of laboratory results for immediate adaptation of medication plan for TB and ARV in rural Transkei[98]. Another very interesting example from the Cape Town region is cell-life, a platform for communication, information and logistical support required to manage HIV/Aids. It is a solution that uses a combination of cellular technology and the Internet to closely monitor adherence of HIV positive patients to antiretroviral (ARV) medication and support community health workers that are visiting AIDS patient in townships[29].

Part II.

The iPath Telemedicine System

Since the mid 1990s, the Department of Pathology at the University of Basel has been researching in the field of telepathology. Early applications included remote frozen section diagnosis with the regional hospital in Samedan[114, 115, 137]. Like many other telepathology systems[112, 133, 151, 156], this application was focused on a point-to-point interaction using specific hardware and software to allow remote collaboration. With fast development of the Internet, a common and relatively inexpensive network has become available almost anywhere in the world. Additionally, modern web browsers have developed to a standard way of accessing information over the Internet, and they provide enough functionality to allow complex applications such as e.g. the remote control of a microscope[16, 62, 122].

In 2000 it was decided to replace the telemicroscopy system at the Department of Pathology in Basel (a point-to-point solution) by an open, web based solution that allows collaboration between any interested pathologists without the need to acquire and install any specific hardware or software. Though some commercial solutions had been available, none met these requirements. The new system was designed as a web based platform on the basis of the PHP programming language and was released to the general public in 2001 in form of the open source iPath telemedicine platform.

The scope of application of the iPath software changed significanlty, when we were approached in 2001 to provide pathology support for a hospital in Solomon Islands. In contrast to the original project, targetting to pathologists in Europe where internet connectivity was not an issue, we were now suddenly confronted with the problems of resource-constrained areas, where infrastructure is minimal and internet connectivity slow and often unreliable. While some projects tried to address this problem by developing solutions that target specifically at low resource settings[49, 140] we decided to integrate the collaborations with partners in developing countries into the same framework that we had started to develop for our own purpose. Allthough this approach was technically more challenging, the decision was based on the assumption that the most limiting factor will not be technology itself but the availability of volunteering medical specialists. Our intention was to lower the threshold for recruiting new specialists by developing a comprehensive and functional platform that lends itself as a useful tool also for collaboration within groups of specialists.

The following chapters will give an overview of design of iPath and on possible applications. Chapter 5 summarises the functionality and architecture of the iPath telemedicine platform, chapter 6 analyses the technical details of the telemicroscopy module, chapter 7 will describe the major applications of the server in Basel and chapter 8 illustrates some examples from developing countries in more detail.

This part is focusing on the methodological aspects of the software infrastructure developed and refined throughout my PhD thesis. The specific methods of the concrete telemedicine applications such as laboratory equipment and image capturing devices are provided separately by each chapter in part III.

5. Description of the iPath Telemedicine Platform

5.1. Technical Background of the iPath-Server

During the scope of the project a web-based telemedicine platform was developed. The platform was named iPath – Internet Pathology Suite – in the initial phase when the applications where exclusively from the field of pathology. In the first version, iPath was basically a set of web dynamic web pages written in the PHP language[1] that had grown out of some preliminary projects at the Department of Pathology. This original version was intended as a proof of concept. Due to its user-friendliness and its ability to accommodate the needs of doctors in developing countries as well as the needs of specialists in western Europe it found good acceptance.

Unfortunately, the original version was not developed following a clear development model and it did not have a good separation of data and visualisation layer. Furthermore, with the steady increase of users and applications utilising iPath, the original version had become very slow. In early 2005 it was decided to develop a completely new version of iPath. This version 2 of iPath includes now an API (Application Programming Interface) which provides a set of classes and functions to access all data stored in the database. In addition, there is now a well-defined concept for splitting the code into independent modules. This greatly facilitates the development of additional modules without changing any of the core code.

A modular and extensible concept is important to allow other organisations to extend and adapt iPath to their own needs. iPath is released under the General Public License (GPL) as Free and Open Source Software (FOSS or OSS) and is thus available to others for usage as well as for modification under the condition that all modifications and improvements are again made available under GPL. Hence, if iPath lends itself to modification and adaptation by others, there is an increased chance that iPath will continue to be developed on a larger scale and by a broader team of developers even after the termination of this very project.

5.1.1. Basic Functionality of iPath

iPath was designed to facilitate consultations and collaboration between different specialists and health care providers. Essentially, iPath offers a collaboration environment for groups of people

---

[1] http:/www.php.net

working together. The basic model for collaboration in iPath is that of a virtual community – a closed user group in which participants can present and discuss medical cases and problems. Every community or group has one or more members that act as moderator who can grant other users access to the group.

Within a group users can create and present content such as a case with a clinical description and attach documents such as images, video clips or forms for capturing alpha-numerical data. Once a case is presented inside a group, other group members can add their comment to the case. Comments are essentially text messages that are attached to a case. In a typical consultation the diagnosis of the experts are inserted as comments. In contrast to other data types, comments cannot be altered by users. Once saved a comment cannot be changed. Only system administrators have the possibility to delete a specific comment. This is important to ensure a certain carefulness from the side of the experts and to make the process of finding a diagnosis transparent. In contrast to email based consultations or video conferences, iPath records the whole diagnostic process and makes it transparently available to all group members.

5.1.2. Application Layers

The new iPath-2 is built in a modular way to ensure that it can easily be extended and adapted to new situations. At the core of iPath is a relational database storing all alpha-numerical information. iPath uses a database abstraction layer (adodb[2]) to ensure compatibility with different databases. At present the open source RDBMS MySQL[3] and PostgreSQL[4] are supported (c.f. fig.5.1).

Most data stored in the database can be accessed using an object oriented API. This API consists of a set of classes which represent users, groups, cases, images, annotations, etc. All classes are derived from a base class (ipath_Object) and the API is extensible in a modular way so that new classes for new data objects can be dynamically added without the necessity to change anything in the core code base. Besides access to data, the API also includes abstraction for general functionality such as registering events (e.g. new case received) with specific alerting methods (e.g. sending an email). A partial documentation of the API is available on the iPath website[5].

5.1.3. The Data Model

iPath is storing four basic types of data objects: users, groups, data objects (cases, images, etc.) and annotations. Technically the iPath API offers a base class (ipath_Object) from which all data classes are derived. This base class takes care of storing and retrieving data from the database and is handling access control.

---

[2] http://adodb.sourceforge.net
[3] http://www.mysql.com
[4] http://www.postgresql.org/
[5] http://ipath.ch/

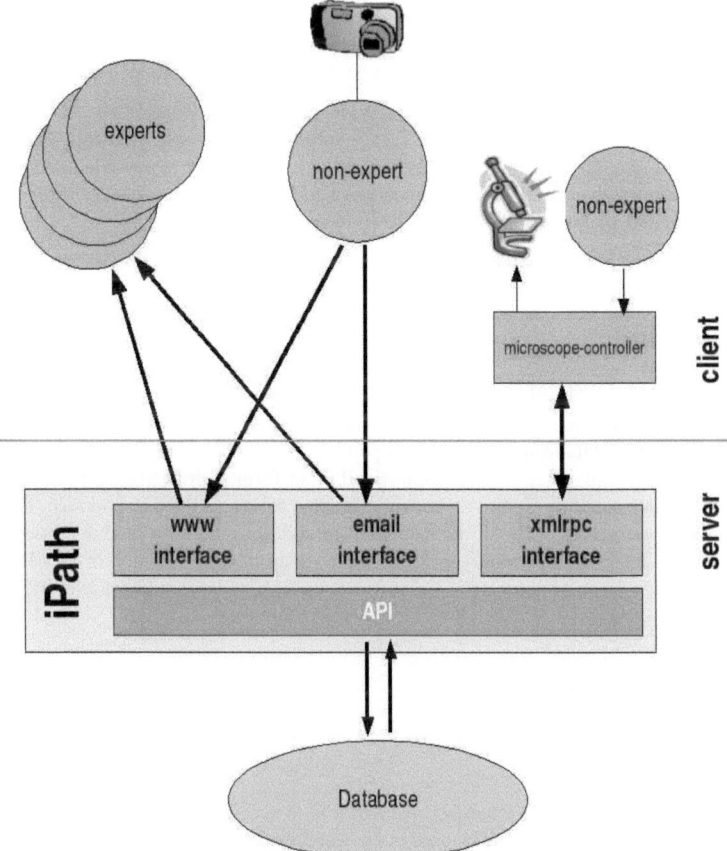

Figure 5.1.: Illustration of the application layers of iPath. At the core is a relational database. User access to the database is provided through different application interfaces (web, email, telemicroscopy). All these interfaces use the iPath API, which provides a set of classes to access data stored in the database.

For the "data objects", only inormation about the relation between objects are stored as fields in the relational database. The medical content is encoded as XML (eXtensible Markup Language) and is stored in the database as a text field. This guarantees a maximum flexibility for creating new data types for representing medical data and findings without the necessity to change the underlying database scheme. The visualisation of medical data on the web interface is done through XSL (eXtensible Stylesheet Language) by applying an XSL stylesheet on the data XML. Thus, new data objects can be created basically by writing a new XSL stylesheet.

Additionally, it is possible to collate several XML objects into a larger XML file. This makes it easy to recursively concatenate all objects representing a medical case (case data, images, folders with images) into one XML file that can be exported. Similarly it is possible to concatenate all cases within a group with all their data into a large XML file which can then be further processed for statistics and analyses using standard XML tools.

All data is stored in UTF-8 character encoding to ensure data input in non-European languages.

Custom Forms

A peculiar data object is the "custom forms" module. This module was developed to provide a very easy way of creating forms for capturing and storing arbitrary alpha-numerical data in iPath. A plain HTML form is used as template for the graphical representation of data. The custom forms module of iPath can automatically transform such an HTML form into an XSL stylesheet. iPath can then use this stylesheet to display a form on the web interface. Data entered by the user is stored as XML in the database. In addition, the form module allows to export data from all forms of a certain type for a whole group in form of a text or HTML table that can be opened with a standard statistical software package or a spreadsheet (e.g. MS Excel) for statistical analysis.

5.1.4. Visualisation

Web Interface

On top of the API there are a number of interfaces to access the data stored in iPath (fig.5.1). The most prominent and frequently used is the web interface that allows access to the data over the web. The web interface offers the most complete set of functionality and is the only way to access the administrative functions. For transforming medical data stored in the data objects into HTML, iPath uses XSL stylesheets which transform the XML coded data into HTML – a procedure termed XSL transformations (XSLT). Other parts of the user interface, mainly administrative pages and settings, are directly rendered as HTML. The advantage of using XSLT is that the stylesheets (XSL) and data representations (XML) are completely independent of programming language and allow utilisation and manipulation of data by external processes written in other programming languages (e.g. Java).

Multi-Lingual User Interface

The user interface is available in multiple languages. A module for on-line translation of the user interface is included in iPath. The translated strings are stored in an XML file which is included by the XSLT process when data objects are rendered. Additionally, the XML files with the translated strings can easily be copied from one server to another, thus translations can be prepared on a test instance and then copied to a productive environment when they are finished. At the moment translations exist for English, German, Spanish, French, Italian, Ukrainian, Lithuanian, Russian, Georgian and Rumantsch. Of course, users can enter content like case descriptions or comments in any language which is supported by the UTF-8 standard.

Email Interface

Besides the web interface, iPath offers an email interface which makes a limited subset of the functionality available over ordinary email. A registered user can submit consultations via email and iPath will import it into the database so that the consultants can access it over the web. The email interface also offers users the possibility to subscribe to email notifications on specific events. Alerts can be defined specifically for each group on iPath. Optionally, the emailed alerts include a full text report with case description and comments of other group members. These reports also contain specially coded "mailto:" links which can be used to trigger a replying email through which the user can 1) retrieve the images for that case or 2) add a comment to the case or 3) if permitted, even add another image to that case by plain email. These emails contain a control hash that identifies the action to be triggered by this email and iPath will securely import data from such emails and add it automatically to the appropriate object in the database.

In contrast to other telemedicine systems developed especially for developing countries that use email for data transmission[49, 169], iPath is a full fledged web application that can handle large amounts of users, organised in many different virtual communities, and large amount of data (100'000 or more images). It is not routing emails to the most appropriate specialist, but imports all cases and comments submitted by email into the same database that can also be accessed over the web interface. Users that have access to email only can still actively participate in the same virtual communities that other users access over the web. While a physician from a remote hospital may chose email to submit cases and receive comments, the consultants in the same group can access the full functionality available – for example, duty plan, drawing applet for annotations in images, web-meetings for the on-line discussion of cases, etc.

An optional module for importing email from the TeleMedMail[49] application is available.

XMLRPC

While the web and email interfaces are intended for human interaction with the database, iPath also offers an interface for interaction with other computer applications. This interface is based on XMLRPC – remote procedure calls (RPC) encoded in XML. This interface offers an API for other modules on iPath to publish certain functions as remote procedure calls accessible to applications running on a distant computer and which can be written in any programming language that support XMLRPC (e.g. JAVA or python). The XMLRPC interface is used e.g. by the telemicroscopy module and also by some specific application to import data into iPath. For example, there is an external application (ImageDrop) which allows importing large amounts of images by drag and drop (current web browser unfortunately do not offer drag and drop functionality).

5.1.5. Modular Design

To encourage further development and adaptation of iPath to other needs, the new version is designed in a very modular way. Almost all functionality is encapsulated into concise modules. The most prominent modules are the content modules that contain the data objects like for example case, image, file, folder, video clip, drawing, etc. Another type of modules are the work flow modules which allow users to simplify a certain task like email alerts or the virtual institute module with its duty plan.

There is a basic set of modules which are essential for the functioning of iPath. These modules are included with the basic distribution of iPath. Other modules that do not include any essential functionality can be optionally enabled through the administrative interface. In addition to the modules bundled with the base distribution, there is a growing number of additional modules which can be added to an iPath installation on an individual basis.

Enabling Modules Group-specifically

One of the most important design considerations for iPath was to keep its user interface as simple as possible. While it is nice to have a lot of optional functionality, medical specialists are often rather deterred by too many buttons and options. In order to provide unrestricted extensibility for some users whilst keeping the user interface simple for others, it is possible to enable certain modules only specifically for certain groups.

5.1.6. Security

All users must first register a user account on the server by filling out a on-line form. When the registration form is saved, iPath will create a user account but mark it as inactive. An activation email is sent to the email address specified by the user. This email contains a special link through which the user must then activate the newly created account by logging in with the password chosen on registration. This procedure ensures that registration is only possible with a correct email address and that the user can receive email on this address.

There are three types of users on iPath. *System administrators* are users that are members of a special admin group. They can access a special system administration interface for configuring and monitoring the system and they can edit almost all data, except comments which can be deleted, but not modified. Besides system administrators there are *moderators* who have some administrative permission for the groups that they moderate – most importantly they can grant and revoke access to the group to other users and the can delete unwanted content from the group. One user can be moderator of multiple groups and one group can have multiple moderators. Normal users finally have access to the groups to which they were granted access by a group moderator and they can only edit or delete content which they have created themselves.

To improve security for data transmission we offer access to our server over https (encrypted http). This feature is not part of iPath itself but can be configured on the web server that iPath is installed on. However, it is possible to configure iPath in a way that iPath switches automatically to encrypted mode, if encryption is available on the web server.

All modifications to content objects are logged to ensure that the full history of changes can be reconstructed at any time. Most administrative requests such as user registrations, password reset requests or granting access to a group are logged by the "watchdog" module. This module also logs every attempted access to content to which a user does not have permission. The watchdog logs the IP address of the remote user, the URL which he tried to access as well as all data submitted via GET or POST requests. Other modules can use the watchdog module to register their own errors and other messages. System admins can access the watchdog entries from the system administration interface.

5.1.7. Advanced Functionality

Beyond the basic functionality of a multi-media discussion forum with closed user groups, iPath offers a number of additional functionality designed especially for telemedicine. Most of this functionality is encapsulated into separate modules in order to make future extensions and adaption easily possible. The most important functionality specifically designed for telemedicine is described in this section.

Virtual Institute / Duty Plan

The virtual institute is a module that enables the organisation of diagnostic consultations. The module itself provides the functionality for the organisation of a duty plan which can then be attached to an expert group. In addition it can create a list of all cases that are not yet closed from an arbitrary number of groups. The usage for this module for the organisation of a virtual institute is illustrated in fig.5.2. A description of its application is provided in chapter 9.

Telemicroscopy

The telemicroscopy module allows real-time remote control of a microscope through the iPath-Server. Routing the remote control through an external server rather than a point-to-point connection offers the advantage that the session can be automatically archived into the database on

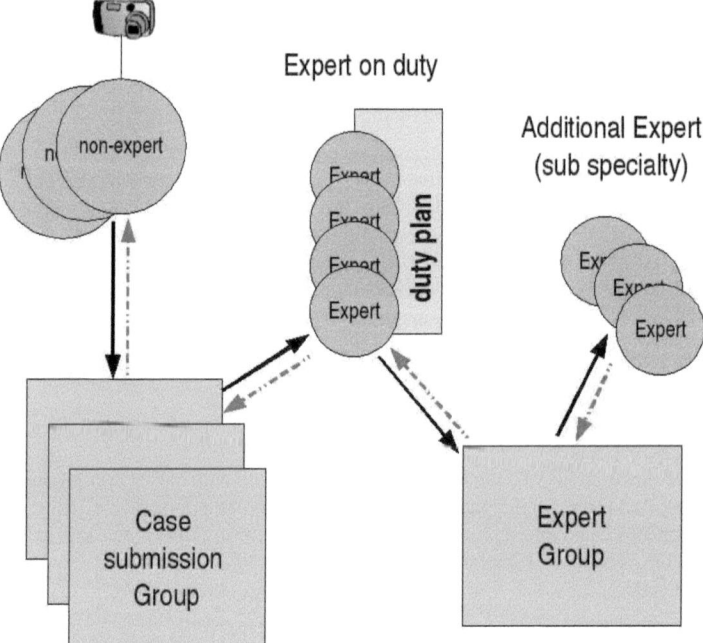

Figure 5.2.: Illustration of the organisation of a virtual institute. The virtual institute consists of one or more case submission groups to which the original consultations are submitted. The expert on duty is automatically notified and can, if possible, write a diagnosis immediately on the original case. If consultation of additional (sub-) specialists is desired, the case can be referred to the expert group. In this case all additional experts are notified and can contribute their diagnosis and comments. Finally, the expert on duty summarises the opinions of the experts and states this as a "final diagnosis" on the original case.

the server. In addition it eliminates the need to configure firewalls in order to allow access from a workstation outside of a hospital to an microscope which is located inside a hospital. This functionality is described in detail in chapter 6. The functionality of this module has been extended since the publication of this article and now includes support for several other types of microscopes, e.g. Leica DM Family and the Nikon CoolScope.

Data Export and Import

iPath offers several possibilities for exporting and importing object data. iPath offers the possibility to export any data object in its XML representations. If an object export is requested, iPath will create a full XML of the object and all its child objects recursively. This XML is bundled together with binary data files (e.g. images) into a zip file which is passed to the user. Conversely it is possible to re-import such exported zip files into another group or even into another iPath server. The only potential problem can be that on the target instance some optional object modules are not installed and the import will be aborted.

The import/export system is easily extensible and developers can write their own import or export plug-ins. Plug-ins can be made part of an add-on module and then a system administrator can enable them group-specifically. Custom export modules developed at the University of Basel include export of a case as pdf file or the export of all comments or data stored in custom forms.

Web-Meeting and Distributed Presentations

The distributed presentation module allows publication and real-time discussion of a case over Internet. The module offers the synchronised viewing of a case using a text chat and a shared pointer. Cases can be discussed within the closed user group or can also be published so that the audience need not necessarily log in on the server. The synchronisation of text messages, image display and position of the shared pointer is implemented using AJAX technology.

Any object that is stored in iPath can be presented through the distributed presentations module. For large audiences with more then 20 participants it is useful to cache the content of a presentation for faster loading and to decrease the database activity during a presentation. iPath offers the possibility to cache the presentation, in that situation, however, only images can be presented. Other objects such as custom forms, drawings or text slides will not work with a cached presentation.

This module can be used for real-time discussion of a case (web-meeting) as well as for distance presentations. There is an option to combine the distance presentation with an audio streaming server for the live transmission of the voice of the presenter. On our server we have implement audio transmission as low bandwidth MP3 streaming using the open source IceCast2 streaming server software[6]. The distance presentations are useful for distance education and are used by the

---

[6]http://www.icecast.org/

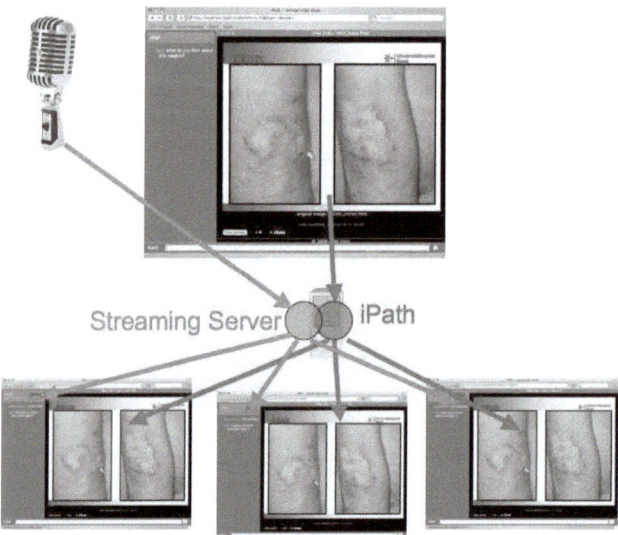

Figure 5.3.: Illustration of a distributed presentations through iPath. The images, chat and shared pointer are distributed directly by the iPath server software. The optional voice data is transmitted with the help of an audio streaming server which was originally developed for Internet radio. The playback of the voice is embedded into the iPath user interface in form of either a JAVA Applet or a RealPlayer Plugin. Thus, for the audience there is no extra software required. The broadcaster of the presentation needs an additional piece of software for recording and broadcasting the voice. Multiple software packages for broadcasting voice are available for most computer systems (Windows, Mac, Linux). With appropriate settings, live audio streaming is possible at relative low bandwidth (4-32KB/s) which makes this also a feasible option for collaboration with developing countries[53].

Swiss Society of Dermatology to transmit presentations from continuous medical education seminars to practitioners who cannot attend the seminar physically. The slides are typically exported from PowerPoint and then imported as cases to iPath. A similar system has been used for the past four years by the French African Telemedicine Network (RAFT) to broadcast presentations to multiple locations in French speaking Africa[53].

Teaching

iPath includes a number of features that allow its application in teaching. In a normal discussion group all material is by default sorted by the date of submission to the server. In a teaching group, the sorting can be predefined by the author of content by providing a "sort number" for every object in a group.

The option to link objects from one group to other groups offers teachers the possibility to separate their pool of material from the actual course that is accessed by the students. Teachers can store all basic material like diagrams, text documents, images or example cases to a specific group that serves as "pool of material". The actual layout of a course is implemented in a separate group to which also the students have access. Original material from the pool is then linked to the specific place within the course. This greatly simplifies the re-use and sharing of material for different courses. While the pool of material can be continuously extended over time, the courses can be easily adapted and re-done according to the teacher's needs.

For practical teaching there are two applications of iPath. 1) Teachers can use e.g. the slide show or the dual projection mode of iPath during the actual lecture as blackboard or slide projector – this can be live in a class room, but also at a distance using the distributed presentations module of iPath. 2) Teachers can grant students access to the information for later review and reference.

5.2. Distribution of iPath

5.2.1. Requirements for Installation of an iPath-Server

iPath is split into several parts. The most important part is the iPath-Server which is a web application written in PHP running on Apache or Microsoft Internet Information Server (IIS). The minimal requirements for installing an iPath instance are:

- Web server: Apache (version 1.3 or 2) or an IIS web server
- PHP: version 5+ with the DOM and XSL modules enabled
- Database: MySQL version 4 or 5 or Postgres version 7 or 8
- Image Processing: preferable ImageMagick or alternatively the gd2 module for PHP. With the gd2 module iPath currently supports only a limited number of image formats.

Optional requirements for all advanced functions:

- The automatic email import requires the php_imap module
- The custom forms module is preferably used with the htmltidy[7] utility for cleaning html form templates and with the iconv (php module) utility for character set conversion when exporting data to excel spreadsheets.

---

[7] htmltidy for automatically cleaning html code: http://tidy.sourceforge.net/

5.2.2. Helper Applications

In addition to the iPath-Server there are several external applications that facilitate working with iPath. The ImageDrop module consists of a Java application that connects to iPath using its XMLRPC interface. Users can drag and drop images and other files from their Desktop to ImageDrop and they will automatically be uploaded and attached to the specified case on an iPath-Server. Similar applications for direct uploading from a TWAIN compatible scanner or from a video camera have been implemented. A more complex external helper is iMic, a Java application that provides an interface between an iPath-Server and a Nikon CoolScope microscope (cf. chapter 6).

5.2.3. Licensing and Availability

iPath is released under GPL, which implies that iPath with the complete source code is freely available to anyone. Anyone may use, redistribute and also modify iPath provided that any modification which are redistributed are again made publicly available under GPL. The source code repository of the iPath server code is hosted on Savannah[8], the software development platform of the Free Software Foundation. The rest of the code which includes all optional server modules as well as the external tools for microscope control and image upload are hosted on sourceforge, another open source software development environment. The code is made available for download in the form of several packages which are all available at http://ipath.sourceforge.net/

A project documentation is available on-line at http://ipath.ch/ and includes the user manuals and installation manuals as well as the API documentation for software developers who want to extend the iPath-Server. Besides the technical documentation there is also a section with documentation of different projects using iPath. The project site is also the web site of the iPath association ("Verein iPath") which was founded in order to promote a sustainable continuation of the projects and services started throughout this initiative.

---

[8]http://savannah.gnu.org/

6. Telemicroscopy by the Internet Revisited

K. Brauchli (1), H. Christen (2), G. Haroske (3), W. Meyer (4), K. D. Kunze (4) and M. Oberholzer (1)

1) Institut für Pathologie, University of Basel, Switzerland
2) Universitätsrechenzentrum, University of Basel, Switzerland
3) Institut für Pathologie, Krankenhaus Dresden-Friedrichstadt, Germany
4) Institut für Pathologie, Technical University of Dresden, Germany

This article has been published in the *Journal of Pathology* 2002; 196: 238-243.[16]

Abstract

This paper reports a fundamentally new concept for Internet-based telemicroscopy. By separating a telemicroscopy application into three tasks - microscope control program, external server, and client application - it is possible to establish a telemicroscopy session between two arbitrary end points on the Internet even if both of the end points are secured by firewall (microscope and client application). The advantages of such a distributed system, compared with the classical point-to-point systems, are discussed. The telemicroscopy system is combined with a telepathology database, which is capable of automatically recording telemicroscopy sessions, allowing a convenient combination of interactive remote microscopy and store and forward telepathology. In addition to remote primary diagnosis, it is easily possible to discuss difficult cases within dedicated user groups, no matter whether images originate from a telemicroscopy session, or are manually entered into the database.

6.1. Introduction

Since the early years of telepathology in the 1980s with specialized hardware and communication links[151], there has been a tremendous development in general purpose (multimedia-) computing as well as in the availability of network links such as the Internet and Integrated Services Digital Network (ISDN). Today, almost any pathologists desktop computer is fast enough for telepathology and most offices are equipped with a fast Internet connection. As yet, however, except for a few point-to-point systems for frozen section diagnosis using specialized hardware[111, 115, 151], there is no widespread use of telepathology.

Besides intra-operative diagnosis, there is an increasing demand for the usage of telepathology for second opinions or scientific consultations[92]. Ideally, a pathologist should be able to connect his/her microscope to the desktop computer in his/her office and use the hospital's Internet connection to share the microscope with any other pathologist with an Internet connection. A concept for telemicroscopy using the Internet and a conventional web browser has recently been introduced[62, 122, 165].

However, a fundamental problem in using the Internet as a network link for telemedicine applications is that transmitted data are basically not secure. As a consequence, most hospitals have secured their internal network (Intranet) with so-called firewalls. These firewalls, however, are a major problem for the use of the Internet for telemedicine, as they prohibit a direct connection between two computers located in two different hospitals[128]. Although it is often not explicitly stated, a direct Internet connection is required by most existing telepathology applications.

In the present article we present a new concept for interactive telemedical applications which allows the use of any Internet connection, secured by any type of firewall, to share a scientific device such as a microscope between several users over the Internet. To illustrate this concept, we will describe our prototype implementation of such a system; however, the main aim of this paper is to demonstrate how it is possible to deal with secured Internet connections as they are present in almost any hospital. The emphasis of this article is therefore clearly on the technical solution for telepathology networks rather than on the diagnostic validity of yet another system. For most pathologists who are trained to make their diagnosis on a real microscope, some time is needed for 'acclimatization' before they are comfortable making diagnoses from digitized images on a computer screen[82].

The complete software referred to in this article and additional technical information is published on-line sourceforge (http://ipath.sourceforge.net).

6.2. Using the Internet for Telemedical Applications

If the Internet is to be used for telemedicine, it is necessary to consider the fact that most users of any telemedicine application will be working in a hospital where Internet connection is almost always secured by firewalls. A firewall is basically a filter that is located between the hospital's

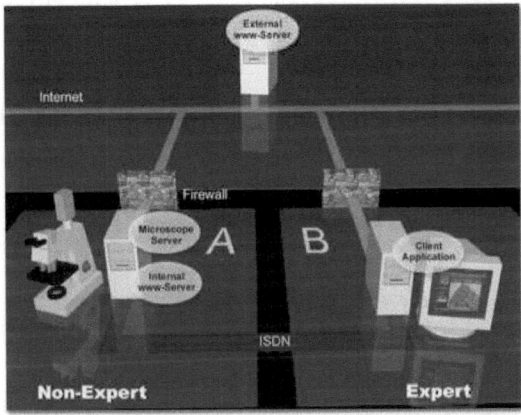

Figure 6.1.: The set-up of the telepathology system described in this paper. For second-opinion consultation, the network connection between the non-expert (hospital A) and the expert (hospital B) is established over the Internet and an external www-server (yellow line). For intra-operative diagnosis, where an open internet connection is not reliable enough, it is possible to run the www-server application on the non-expertâs computer that is connected to the microscope. The network link can be established over a private ISDN line.

intranet and the open Internet. This filter allows only connections from within the hospital to the outside and only for certain networking protocols (such as HTTP and e-mail). It is therefore not possible to connect directly two computers located at two different hospitals, as each attempt to establish a connection would be blocked by the other's firewall.

To allow any use of the Internet, a firewall must at least tolerate outbound connections to receive web pages. Technically, this means that the hypertext transfer protocol (HTTP) is always open for outbound connections. Other protocols such as File Transfer Protocol (FTP) or Telnet are often closed and inbound connections are almost never tolerated. If a telepathology system is to make use of the Internet as it is present on a typical hospital's Intranet, it must therefore only make use of outbound HTTP connections for data transfer. A simple but practicable example of such a system, using common webcam technology (with image upload over FTP), was recently established in the UK[128].

To implement a telemicroscopy system under such restricted conditions, we separate the telemicroscopy application into three tasks, which are implemented as independent modules: a 'microscope control program', the 'client application', and a 'telepathology server', which is located in the Internet and routes data between the microscope control program and client application for an arbitrary number of telepathology sessions (cf. Figure 6.11). In the following section, we give a brief overview of these modules as they are implemented in our prototype system.

6.3. A Way Out: Distributed System for Telemicroscopy

6.3.1. Connecting the Microscope: the Microscope Control Program

The microscope control program is an application that on one side physically controls the microscope and captures images from a camera, while on the other it can communicate with the www-server. To discuss a difficult case with a remote expert, a pathologist starts the microscope control program, which has to be installed on his/her computer. After setting illumination and optionally preparing a map image (an overview which is composed of several images at low magnification) the microscope control program registers the microscope with the www-server and waits for instructions from remote clients. If the local or nonspecialist pathologist has a motorized microscope, the microscope control program will operate it according to instructions from the clients. Without a motorized microscope, the non-expert must operate the microscope manually. The expert pathologist may instruct the non-expert over a conventional telephone connection. To direct the non-expert to a desired field of view, the expert can use the point and click interface as for a motorized microscope and the non-expert can see the desired position on screen and can change the stage position accordingly. With some experience by the non-expert, it is easily possible to do remote microscopy with a conventional microscope.

In addition to translating instructions from the expert into microscope commands, the microscope control program also sends some sort of 'live image' to the server, from where it is retrieved by the clients. As there is no guaranteed bandwidth on the Internet, we have reduced the transmission rate for the 'live image' to one static image every 1 or 2 s, instead of trying to provide a real-time video. We find that such a low frame rate is enough for most telemicroscopy tasks, except maybe adjusting focus from a distance, but this can be done by the non-expert, who is next to the microscope. As these 'live images' can be heavily compressed, it is possible to provide an almost live feeling for remote microscopy with very conservative bandwidth needs: a single ISDN channel (64 kbit/s) is enough to perform on-line telemicroscopy. However, if bandwidth and fast computers are available, it should be easily possible to use a compressed video stream instead.

At present, our microscope control program software is running under Microsoft Windows NT/95 and can be used with any 'Video for Windows' (VFW) compatible video input card, and it can optionally control a Zeiss MCU-26 stage controller. Ports to other hardware are in preparation. Support for high-quality frame grabber cards (Matrox Meteor and IC-PCI) is almost completed and support for other microscopes (Olympus and Leica) and OS (Linux) is being worked on. A motorized microscope is not required; a conventional microscope operated by hand, although slightly slower, is sufficient for most tasks.

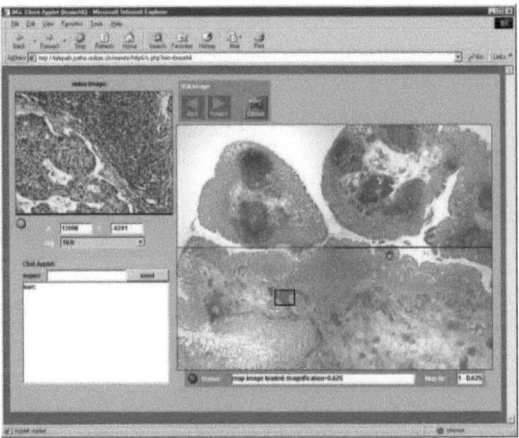

Figure 6.2.: The client application is implemented as a JAVA enhanced web page, which can be used with a conventional web browser. The "live" video image in the top left corner always shows the actual view from the microscope, automatically updated every 1-2 s. On the right, the last captured image is displayed. With a motorized microscope, the actual position of the live image is always outlined by the black rectangle. With a conventional microscope, the rectangle can be used by the expert to direct the manual operation of the microscope.

6.3.2. Using the Microscope at a Distance: the Client Web Page

The client application is implemented as a JAVA enhanced web page, which has proven to be suitable for a telemicroscopy client[62, 122, 165], as it can be used on any computer platform equipped with a conventional web browser such as Netscape Navigator or Internet Explorer. Figure 6.2 illustrates how the microscope is represented to the remote expert. A small 'live image' (top left, updated regularly according to the present bandwidth) always shows the current view under the microscope. Single high-quality images (true colour VGA) can be captured and are then displayed on the right side. Images are stored and transported in standard JPEG format. The compression rate can be set dynamically, which allows the users to make a reasonable trade-off between speed and image quality. If the magnification is increased or the stage moved, the new position is indicated on the captured image with a rectangle (cf. Figure 6.2). The microscope can be directed to a new field of interest by clicking at the desired position in the captured image.

As the application can store several captured images, it is always possible for the expert to scroll back to earlier locations and continue the investigation from there. In addition, every session is recorded by the telepathology server, so that it can be reviewed even when the microscope is no longer on-line, to allow archiving and later discussion of difficult cases (cf. Figure 6.3).

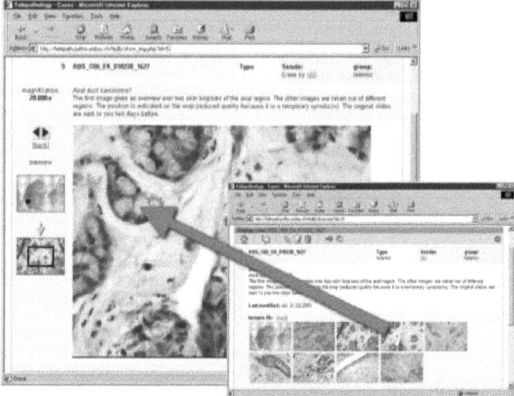

Figure 6.3.: Every telepathology session can be recorded by the www-server, including all captured images with their exact position and magnification. Hence, the complete session can be reviewed or re-discussed at any time without access to the microscope. The real slide must be put back under a microscope only if other areas of the slide are required. In this case, the database will record further images along with those already existing.

6.3.3. The Telepathology Server

As in most cases, due to restrictive firewalls, it is not possible to establish a direct connection between client application and the microscope control program, we introduce an external server to route data between the microscope and clients. Basically, the telepathology server is a web server with support for active web pages (PHP and JAVA) and a collection of CGI scripts for routing data between the microscope and client.

If a telemicroscopy session is to be established, the microscope control program first contacts the telepathology server, authenticates itself and sends the state of the microscope (position, magnification, etc.), and then waits for the next command to be executed. If a client requests some action from the microscope, the telepathology server forwards this request to the listening microscope control program, waits for the answer, and distributes the answer (e.g. a captured image) back to the client. Additionally, the current state of the microscope is stored on the server, so that it can be distributed to all clients without calling back every time to the microscope control program.

As a side-effect of such a triangular architecture, there is always a delay compared with a direct point-to-point connection; the distributed set-up takes about 1-3 s more to complete a request than a direct connection. Compared with the advantage of being able to use almost any kind of Internet connection on both sides, this is in our opinion a minimal disadvantage.

Besides routing information and providing access control, the telepathology server can additionally record every session into a database, including all images, with their exact position and magnification (cf. Figure 6.3). This has proven to be very valuable: sessions can be reviewed and discussed at any time without accessing the microscope; images that have been important for a diagnosis can be downloaded and stored along with other patient information after the interactive session is finished (from all involved partners); and cases of special interest can be published on the telepathology server to a general audience by one click of the mouse.

As no Internet server can be fully secured, it should again be stressed here that any data that could be used to identify patients should not be stored on the Internet. If sensitive data are needed along with the images, a simple but reliable solution is the use of a different, secure transport medium, such as spoken language over the telephone or encrypted e-mail (encrypted with PGP[17]).

6.4. Discussion

We have shown how to design a telepathology system that uses the Internet as a network link, but is not limited in its use by the restrictive firewalls present in most hospitals. The distributed approach to accomplish this task has many more advantages, discussed below. Lack of interoperability has been identified as a major drawback of existing telepathology systems[159]. One way of overcoming this problem, an open standard, is almost implicit in a distributed system.

6.4.1. An Open Standard

The distribution of process clearly separates the telepathology application into several distinct modules. Provided that the interfaces between these modules are standardized and published, single modules can be implemented or adapted to specific hardware independently from the rest of the system. Improvements to the server, database or client applets are immediately available to every expert who is using a connection via the server, while the hardware-specific microscope control programs do not need any change at all.

Additionally, the development of new components is no longer restricted to a single vendor or any central authority. At present, development on the telepathology server software will be continued as an academic project with additional partners. The emerging software will be released under the existing open source licence. On the other hand, it would be easy to imagine that microscope vendors could develop and support commercial versions of the hardware-specific microscope control programs.

As an example of such an open standard, we have published the complete communication interfaces of our prototype system on-line (see below).

6.4.2. Use for Frozen Section Diagnosis

While a range of commercial applications exist especially for frozen section diagnosis, it is perfectly possible to use such a distributed, open system, provided some attention is paid to security and reliability. For example, the telepathology server module can be installed on the same computer as the microscope control program, or anywhere within a hospital's Intranet. Network connection to the expert is established over the Intranet (in-house diagnosis) or over ISDN, so bandwidth need and reliability can be guaranteed, as well as privacy of exchanged data (cf. blue route in Figure 6.1). By dynamically adapting compression and the transmission rate of images, the system described is functional under very limited bandwidth conditions (64 kbit/s). If external help is needed, the same microscope can be additionally connected to an external telepathology server over the Internet and thus an expert can be consulted immediately.

A small pilot study on routine material demonstrated the same diagnostic performance as desired in routine frozen section diagnosis; neither false-positive nor false-negative tumour diagnoses and a deferral rate below 10% [65]. Compared with a commercial telemicroscopy system, connected by three multiplexed ISDN lines with a robotic microscope, which was used for remote frozen section diagnosis on the same cases, the median time needed was 8 min versus 6 min for telemicroscopy via the Internet versus the commercial system, respectively. It was found that a surgical pathologist does not need special training in the use of such a system, if already familiar with usual desktop PCs.

6.4.3. Combining Dynamic Telemicroscopy with 'Store and Forward' Consultation

Besides on-line telemicroscopy, discussion of difficult cases in dedicated user groups is a very important application of telepathology[92]. The distributed system described above allows a very convenient combination of interactive telemicroscopy and 'store and forward' consultations in the very same application. While all telemicroscopy sessions are automatically recorded to the database, there is also the possibility to enter images manually. The advantage of such a flexible solution can be illustrated with some examples. Firstly, images from different sources, such as a gross specimen or even radiology, can be stored along with the images from an on-line consultation. Secondly, a consultation may be started with a set of images provided by the non-expert, but if the expert wants more information and a telemicroscopy session is then conducted, the images captured during that session are stored automatically, along with those provided by the non-expert beforehand. The presentation and handling of the database are always the same, no matter where the images originated.

6.4.4. Future Directions

A first improvement planned for the future is to incorporate a range of collaborative tools that will improve the discussion within a user group. The first step, which is almost completed, is a text-based chat function that allows on-line discussion between more than two partners. There are some additional advantages of chat-based communication over voice telephone: the communication can be easily stored in text form and there are fewer communication problems between partners with different native languages when using a common language (English) in written form, instead of oral communication (R. Weinstein, oral communication). Due to its modular concept, the system can be easily extended with other collaborative tools, such as a shared pointer or a whiteboard.

Additionally, there are plans to make the telepathology server inter-operable with other existing standards such as DICOM, whereby other image sources (CT, MRI) could be used to feed the telepathology database directly.

6.5. Conclusion

While in some special fields telepathology has become reasonably accepted by a variety of users, a general, widely accepted framework is not yet in sight. Most existing systems are conceptually targeted to restricted fields of application and are therefore not very useful as general tools to most pathologists. Although the presented solution is only a prototype, we hope that its concepts and the ideas presented in this article demonstrate how a general framework for telepathology that includes a broad field of telepathological applications, from intra-operative frozen section diagnosis to inter-continental second-opinion consultations, can be implemented using technology that is available today. We believe that only an open, modular framework will help to overcome the obstacles that still prevent the more widespread use of telepathology.

The complete software described in this article is freely available as open source software. The software and more detailed information on the project, especially the communications standard implemented, can be found at http://ipath.sourceforge.net

Acknowledgement

We would like to thank Dr Peter Furness for his critical and most valuable comments on our manuscript.

7. Applications of the iPath Server at the University of Basel

Since 2001, the Department of Pathology in Basel has been using the iPath platform to operate a global telepathology network (http://telepath.patho.unibas.ch/). Today this network has over 1200 registered users organised in over 100 discussion or working groups.

In April 2006 these various working groups contained over 8'000 cases with a total of over 55'000 images and 3'500 other file attachments. Over 14'500 diagnostic comments had been already submitted. Table 7.1 gives an overview of the activity on this server. From over 1200 registered users, 906 had made use of their account and had logged into the server at least once after registration. On average a user had logged in 47 times, had presented 9 cases, provided 19 comments and visited 59 cases. There is of course a huge variance between the different users. While 237 users had not visited anything but test or demo cases, the most active user had visited 4146 different cases. 262 users had presented at least one case and 347 had written at least one comment. The most active users had presented 1302 cases and written 2137 comments. The bottom row illustrates the number of user who had logged in more than 10 times, who had submitted 10 or more cases or comments and who had visited more than 10 cases. Obviously there is a substantial number of "passive" users that do visit the cases but do not write any comments or present their own cases.

7.1. The iPath-Server at the University of Basel

The very first iPath server at the University of Basel based on a old Desktop PC (Pentium II), had been installed by the end of 2000 and served mainly as proof of concept. With the launch of the projects on Solomon Islands and in Cambodia in 2001 the amount of consultations started to grow and it was decided to migrate the project to a proper server hardware. In early 2002 iPath was migrated to a Fujitsu Siemens Primergy Server (rack server with a 1GHz Pentium III CPU and 36GB SCSI hard-drive) running on RedHat Linux (version 7.2) operating system. In 2005 the storage capacity was extended to 145GB (2 hard disks in mirroring RAID configuration) and the operating system was changed to a Debian based Linux.

Table 7.1.: In April 2006 the ipath Server at the University of Basel had 1283 active users and 769 inactive user, which had never activated their account or had not logged in for more than one year. This table includes the data from all active user that had logged in more than once. Administrative users and test cases are excluded from this table.
Column one shows the number of logins, column 2 the number of cases submitted, column 3 the number of comments contributed and the last column shows how many times a case was visited.
The two rows at the bottom show the number of users who have e.g. submitted more than one case or 10 or more cases respectively.

	login	cases	comments	case visits
avg	47	9	16	59
max	2794	1302	2137	4146
sum	42664	7808	14130	53293
nr. users > 1	906	262	347	669
nr. users >= 10	392	72	119	371

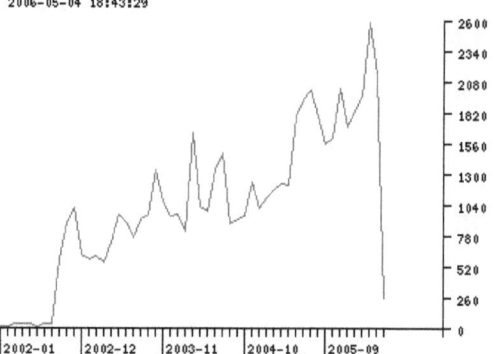

Figure 7.1.: Login statistics for the iPath-server at the University of Basel. The y-axis illustrates the number of monthly user sessions. Valid figures are only available since August 2002.

54

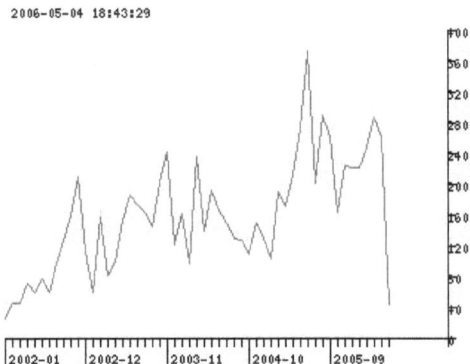

Figure 7.2.: Monthly case submission to the iPath-server at the University of Basel.

7.2. Overview of Applications and Groups

On April 1st 2006, 2056 user accounts were registered on the iPath server at the University of Basel. Out of these 2056 users, 1222 have logged in more than once in the last 6 months and can thus be considered as active. These users were organised in 67 active groups – counting only groups that were not set up for testing and which contain more than 10 cases. In total there were 154 groups, but many of them were created for some test purpose or never became active.

Most of these groups are not organised by the core iPath team at the University of Basel. During the project there were people from all continents approaching us with new ideas for using iPath. For many we granted access to our server and let them organise their own groups. As with many new technologies, there was often an initial boost of interest but afterwards most groups never achieved a critical mass to get an interesting discussion going and later ceased to exist. Other groups existed for years and exhibited continuous activity. To give an overview of what kind of applications the iPath server is being used for, I will give a brief description of the most active groups. Raw data for this overview are listed at the end of this chapter.

7.2.1. Groups with over 100 Cases

Brustzentrum Dresden (1375 cases)

Virtual community of the breast tumour competence centre of the region of Dresden, Germany. Currently three hospitals and three community clinics are participating in this breast tumour centre. The working group on iPath is used for the clinical documentation of all breast tumour patients treated at any of the participating institutions. The group uses a set of 15 forms to capture data from different specialities including oncology, pathology and radiology. Different

55

examinations are performed in different locations. None of the participants had ready access to all the data (radiology, histology, pre- and post-operational examinations, follow-ups, etc.) necessary to perform the quality assurance statistics that they are required by the government. Since the institutions do not share a common medical IT platform, the collection of all data on all common patients has to be organised over an external application to which all participants have access. The solution was based on iPath and allows each institution to insert and modify all data generated at their side. They also have ready access to all other data. The data collection is utilised for the interdisciplinary tumour board meetings as well as for extracting the indicators for reports and quality assurance statistics. At the moment the calculation of these statistics is done with an external database at one of the participants site, but it is planned to integrate this as an additional module for iPath using the XQuery language and XSLT language for generating report templates.

Sihanouk Hospital Center of Hope, Cambodia (1296 cases)

This working group is formed by the Sihanouk Hospital Centre of Hope in Phnom Penh and pathologists from Germany, Switzerland and Thailand to provide diagnostic services in histo- and cytopathology for the hospital in Cambodia. A detailed evaluation of this application is published as a working paper in chapter 11.

South Pacific (499 cases)

Virtual Community formed by the National Referral Hospital in Honiara, Solomon Islands and a group of pathologists providing diagnostic services in histopathology for the Hospital in Honiara. A detailed description follows in chapters 9 and 10.

Oncology Centre Lörrach (475 cases)

The oncology centre Lörrach is used for supplementing tumour board meeting held at the regional hospital in Lörrach, Germany. Participating partners in the tumour board meeting include the pathologists from the University Hospital in Basel, radiologists from the hospital in Rheinfelden, Germany, and from private practises in the region, the oncologists from Lörrach as well as several external specialists. If the clinicians in Lörrach want to discuss a case in the bi-weekly tumour board meeting they create a case in the discussion group on iPath and they announce this to the partners who will then attach their material (mainly radiology and pathology images) to the cases. During the actual meeting, all material is presented directly from iPath.

This has the advantage that all material is available before and after the meetings for those who are interested in the case and want to review the presented material. Additionally, using the dual projection module, cases can easily be compared with each other, for example, with a former biopsy if it is already in the database. Furthermore, additional material such as schemata or papers can be added to the case to illustrate the findings. Another advantage is that these presentations stored on iPath are building a very interesting archive available to all clinicians participating in the project. And finally, since the material is already collected on-line, it is very easy to make it available to an external expert for a second opinion.

All cases are stored in an anonymous form without a direct patient identifier. Patient identifications are communicated over a separate channel.

Bangladesh (346 cases)

A discussion group started in March 2002 by a pathologist from Dhaka, Bangladesh, to discuss complicated and unclear cases with colleagues from the UK and from South Africa. Over time, the group of participating pathologists has grown and now includes experts from almost all continents.

Feldstudie Telepathologie (325 cases)

This group was used for a review study of the German Pathology Association to collect material from telepathology consultation in anonymous form in order to asses the diagnostic quality of telepathology consultations done with various real-time telepathology systems. From February 2002 to April 2004 over 300 cases were collected. However, it is unknown to us if the material collected in this way has ever been evaluated. A peculiarity of this group was that it was configured as an "anonymous group" which ensures that usernames of the submitters of case and comments are not visible to group members. Only the group moderator can see the identity of the contributions.

Basel Atlas Daicker (247 cases)

Group used for the edition of the illustrations for "Stereoatlas of Ophthalmic Pathology" – a book by P. Meyer (Basel, Switzerland) and K. Löffler (Bonn, Germany) published by Karger Verlag, Basel. From the whole compilation of the totally over 12'000 stereoscopic slides of interesting and rare eye diseases, the editors selected 246 pairs to make them available to interested health professionals. iPath was used for the selection process as the two author and the publisher were working in geographically different locations. Additionally, the content of the CD-ROM accompanying the book was directly produced from the database of our iPath server.

Laos Vientiane (239 cases)

Diagnostic support for pathologists from the National University of Lao in Vientiane, Laos. Pathologists from the University of Calgary in Canada are forming the team of experts. They are supported by volunteering pathologists from Germany. This group started in 2003.

Histopathology Forum (225 cases)

In the early days of iPath, most applications were either defined groups of experts providing diagnostic consultations for a defined partner or discussion groups for case discussions within existing working groups. Over time, more and more independent pathologists from all over the world started to register in our server. Since these pathologists did not belong to any of the existing user groups, a new, relatively open group named "Histopathology Forum" was started in December 2004. In contrast to the closed user groups, where access is only granted by the true group moderators, any person interested in participating in the Histopathology Forum will be granted access directly by the system administrators.

Since December 2004 over 240 pathologists have joined this group. Cases from all over the world are frequently presented and debated sometimes in lengthy discussions with up to 14 comments per case. With an average of 4.75 comments per case this group shows one of the highest commenting frequencies. In addition to the comments posted to the groups itself, some cases were discussed further in a closed expert group such as the hematopathology group. On average, a case in the Histopathology forum is visited 88 times which is another sign of the high activity in this group.

Basel Ophthalmology (203 cases)

Closed Discussion group of ophthalmologists from the University of Basel. Besides consultations, this group was also frequently used for the selection of illustrations for books and the editing of the accompanying figure legends.

Telecytology Basel (168 cases) and Telecytology Honiara (35 cases)

While tele-histopathology has been studied quite well, there is very few literature available over the feasibility of tele-cytopathology. However, cytology is often much easier and cheaper in preparation and can give very valuable diagnostic results for certain diseases like, for example, cervical carcinoma which is relatively prominent on Solomon Islands. For Solomon Islands no published data is available but in a study carried out in the neighbouring Vanuatu which has a pre-dominantly Melanesian population like the Solomon Islands: Pakosy et al.[120] found that cervical carcinoma was the most frequent cancer of the female population (25.5% of all carcinoma).

A major problem with cytology is a lack of qualified cytologists in developing countries. However, a well-trained technician can screen cytology slides very effectively and for most cytology diagnosis a qualified cytologist would be mainly necessary for quality control, continuous training and for difficult specimen. Thus, theoretically, cytology could be a prime application for telemedicine. In a joint study of the National Referral Hospital in Honiara, the Ministry of Health of Solomon Islands and the Department of Pathology of the University of Basel the feasibility and diagnostic quality of tele-cytology under conditions of a typical hospital in a resource-constrained area is being evaluated.

As a control study, 168 randomly selected routine cytology cases from the laboratory of the Department of Pathology in Basel have been photographed by a cytologists and presented on iPath to a group of pathologists: three specialists for cytology and one specialist for histopathology. The four specialists diagnosed each case and entered their diagnosis to iPath without being able to read the others diagnosis. Using the option "hide comments" on each case, the diagnosis of the other colleagues only becomes visible after a user has entered his own diagnosis. A preliminary evaluation of the results of the control study is included below.

Iran Pathology Group (163 cases)

The Iran Pathology group includes two different applications. In March 2002 a pathologist from the Tehran University of Medical Sciences started to introduce IHC (Immunohistochemistry) staining in Iran. IHC uses antibodies to specifically mark certain proteins in the tissue. In order to ensure the quality of the IHC method and to ensure that the new diagnostic tool was used in line with standardised WHO classification, the Iranian pathologists used iPath to discuss certain cases with pathologists from Europe with long-term experience in IHC staining.

The same discussion group on iPath was later used in a study carried out by a pathologist from the Department of Pathology at the University of Kerman, Iran, and colleagues from the Institute of Pathology at the Charitée University in Berlin, Germany. The study compared the diagnostic outcome of the UICC telepathology centre at the Charitée in Berlin and the iPath Server of the University of Basel [19, 101].

(The group ceased to be operational in 2004 due to unknown reasons. Perhaps with the successful introduction of IHC staining and the publication of the results from the telepathology study, the interest in tele-consultations have disappeared.)

Radiology Lörrach (162 cases)

Archive of interesting radiology cases by a radiologist in Lörrach.

Ukrainian Swiss Perinatal Health Programme (159 cases)

Within the framework of the Ukrainian Swiss perinatal health programme organised by the Swiss Center of International Health, a telemedicine component was established. This telemedicine component is based on iPath and tries to foster collaboration between specialists in Ukraine and in Western Europe in the field of neonatology and obstetrics. Most images submitted are ultra sound and x-ray images and also photographs of patients.

The Ukrainian Swiss Perinatal Health Programme has now established its own server in Kiev, Ukraine and the data will be moved from the server at the University of Basel to that server. A detailed description is given in chapter 13.

AG Knochentumoren (140 cases)

The very first group using iPath since 2001 is the German bone tumour working group (AG Knochentumoren). They use their discussion group on iPath mainly for the preparation of their biannual meetings, where difficult and interesting cases are discussed with the goal of finding a consensus diagnosis. To prepare these meetings, most cases are published in iPath and discussed there. This group uses the feature of "hidden comments" that hides comments of others for a user until he or she has entered his or her own diagnosis. Looking at the activity per single case this is the most active group: a case receives on average 7.1 comments and is visited 118.6 times. The group is also infrequently used for "ad hoc"-consultations.

Some of the members of this groups are also active as bone tumour specialists for various other groups.

AG Mamma- und Gynäkopathology der SGP (120 cases)

Similarly to the "AG Knochentumoren" the working group on mamma- and gynaecopathology of the Swiss Society of Pathology has started a discussion group on iPath for preparation of their consensus meetings.

Expertgroup Pathology Basel (119 cases)

This group is used for the organisation of the "Virtual Institute"[21, 18] that is available for some of the hospitals in developing countries that are frequently submitting cases (Solomons, Cambodia, Laos and Bangladesh). A virtual institute technically consists of a separate expert group in which all participating specialists are organised. The expert group has a duty plan so that for every week it is clearly defined which pathologist is "on duty". Several other groups may be linked to the expert group and the expert on duty has access to a special case list on which all new cases from any of the linked groups are listed. If a diagnosis is easily possible, the expert on duty will write a "final diagnosis" and thus close the case – an email with the diagnosis is automatically sent to the case submitter. For more complex cases, the expert on duty can make a referral to the expert group and then all members of the expert group are notified. The case will now be discussed within the expert group and after a few days, the expert on duty is obliged to summarise the discussion and publish this summary as final diagnosis back to the original case (cf. fig.5.2).

Optionally, it is also possible to grant access to the internal discussion to the submitting doctor. Following the discussion of the experts can be a very useful learning experience for the submitting physician, however, it may also be confusing if different possibilities are suggested and discussed. Thus, some submitting physicians prefer to only receive the summary of the expert on call. The choice should be with the submitting physician.

7.2.2. Other Interesting Groups

Beside the very active groups with over 100 cases there are several interesting groups worth mentioning.

Zürich Neuropathology

For one year the neuropathology specialists of the Department of Pathology in Basel is on a research sabbatical in Zürich and thus there is no neuropathology specialist in Basel for dealing with complicated intra-operative frozen section diagnosis that are sometimes requested by the neurosurgery ward of the University Hospital. In order to overcome this problem, difficult neuropathology specimen are presented to the specialist in Zürich using a fully motorised microscope (Nikon CoolScope) and the remote microscopy module on the iPath server. Using this technology allows the specialist in Zürich to review a slide in Basel within minutes.

Ethiopia Pathology (61 cases)

The "Ethiopia Pathology" group is organised by a retired Swiss pathologist, who is currently working for the Tikur Anbessa Hospital (Black Lion's) Hospital in Addis Abeba, Ethiopia. The group is used by this pathologist and his Ethiopian colleagues for second opinion consultations with pathologists from Zurich and Germany. So far 61 cases have been discussed and received on average 2.3 comments. One consultation was published as case report in the Ethiopian Journal of Medicine[132].

Hematopathology (52 Cases)

An expert group for the discussion of hematopathology cases, organised by a retired pathologist from Basel. The group serves for the discussion of interesting cases, but mainly as expert group for the referral of difficult hematopathology cases from other groups. A major problem for organising sub-specialty consultations is that it is difficult to motivate specialists to review cases, especially if the quality of the submitted cases is not very high.

Most users on our iPath server do not have direct access to the hematopathology group. However, if an interesting hematopathology case is presented for example in the Histopathology Forum or from Cambodia, the system administrators will inform a member of the hematopathology group who may then refer the case to the expert group, provided that the submission is of interest and acceptable quality. This triage model is a very efficient tool to organise very specific subspeciality consultations.

Port St. Johns, South Africa

This group was created in 2002 to facilitate the teledermatology consultations between a GP in Port St. Johns, South Africa, and some dermatologists in the public hospitals in Mthatha and East London. This application was initially running on the server at the University of Basel and was later moved to a regional server operated by the Walter Sisulu University (WSU) in Mthatha in June 2003. Since then, all tele-consultations from Port St. Johns have been handled on the server at WSU. A detailed description of these activities is provided in chapter 12.

Norodom Sihanouk Hospital, Cambodia

The Norodom Sihanouk Hospital is a large public hospital in Phnom Penh, Cambodia. It is one of the very few centres that offer reconstructive surgery. The surgeon is using the iPath facility to consult with colleagues in USA, Australia and Switzerland. The idea for the drawing module, which allows a consultant to draw an annotation directly into an image submitted to iPath, originated from this application.

African Dermatology Forum/Tropical Dermatology and Venerology Forum

Occasionally, dermatology consultations from several developing countries are submitted to iPath through one of the pathology consultation groups. The "African Dermatology Forum" was a first attempt to facilitate dermatology consultation. It had received a few cases in the beginning but then ceased to operate. The name implied that the forum was limited to Africa and some users on iPath did not feel comfortable with this. On request from Cambodia a new group was started with a more generic name. Hopefully it will be possible to disseminate the knowledge about this group. To ensure a certain quality of contributions, it is envisaged to link it with the activities of the Swiss Society of Dermatology, which is using iPath for their dermArena project (see next section on "vitual portals").

Basel Pathology Groups

Additionally, there are various groups used by the pathologists in Basel. They are mainly used to disseminate interesting findings to clinicians who submitted the material to the Department of Pathology using the distributed presentations facility provided by iPath. On some occasions, iPath was used to get additional information from the surgeon who submitted the tissue for pathology examination.

7.2.3. Virtual Portals

Several groups have gone a step beyond creating a discussion group only and have actually started their own virtual portal on iPath. A virtual portal is a separate entry to the same database, but having a different layout, front page and often a different default language. Virtual portals are also important to create some sort of corporate identity which is often a highly motivating factor for new members. Most virtual portals consist of more than one discussion group.

Lithuanian Pathology Online (46 cases)

A group of Lithuanian pathologists have started their own virtual portal on our iPath server (http://telemed.ipath.ch/lithuania/). The group has translated iPath into Lithuanian Language. The portal consists of several groups of which the open forum names "Lithuania Pathology Online" is the most active. Plans for the future include a discussion group for providing continuous medical education, mainly in Lithuanian language. A baltic lymphoma registry based on iPath is also discussed.

PathoIndia (43 cases)

Virtual portal of the pathoindia project which brings together pathologists from all over India (http://www.pathoindia.com). For some time the groups was using a discussion group on iPath with a virtual portal (http://telemed.ipath.ch/pathoindia) for their case discussions. However, most active members have later moved to the new Histology Forum and are now using this more international group for the discussion of interesting and complicated cases.

Donetsk Regional Telemedicine Group (39 cases)

A group of physicians from the region of Donetsk, Ukraine, who translated iPath into Ukrainian and who run their own virtual portal on http://telemed.ipath.ch/donetsk. This groups is mainly used for presentations and discussions in the field of gynaecology and obstetrics.

RAFT- Forum (16 cases)

RAFT is the "réseau de télé-enseignement et de télémédecine en Afrique francophone" (French African tele-education and tele-medicine network) organised from Bamako, Mali, and the University of Geneva, Switzerland. Since 2001 the network has been active and the main activity has been broadcasting lectures from Europe to Africa and also from Africa to Europe. For example, the students of tropical medicine in Geneva can use the network to attend lectures from the medical faculty of the University of Dakar, Senegal. These distance lectures are organised by means of the e-Course platform of the University of Geneva[53]. Recently, the network has started to use iPath for tele-consultations and is running a virtual portal on our server, on which the default language is French (http://telemed.ipath.ch/raft/).

DermArena - Suisse DermatoPratique (14 cases)

DermArena is one of the youngest projects on iPath. It is a virtual portal for the Swiss Society of Dermatology used mainly for broadcasting CME seminars, regularly organised by the society. This group utilises the distributed presentation module of iPath. An application associated with DermArena are the virtual case rounds organised be the French-speaking dermatologists in Switzerland using the group "Suisse DermatoPratique". Once a month an interesting or problematic case is presented live using the distributed presentation module within a group of Swiss dermatologists. While the number of cases (14) is not very high, it is interesting to note that the group has by far the highest number of visits per case: on average every case is visited 170 times!

Solomon Islands Telemedicine Network

Since 2001 the National Referral Hospital in Honiara (NRH) has been using iPath for telepathology consultations with external specialists. In 2005 a group of physicians from NRH together with the Ministry of Health has come up with the idea of using iPath to provide consultations to the provincial clinics which are mostly located on other islands. Nurses and doctors from the provincial clinics shall have the possibility to consult with the specialists in Honiara. If there is no appropriate specialist available in Honiara, an external group of specialists for secondary consultation is previewed.

Besides the consultations for provincial clinics, the network will also enable consultations in respect with physical patient referrals from NRH to the St. Vincent's Hospital in Sydney, Australia. Additionally, the Solomon Islands telemedicine network is also planning to use the distributed presentations module to organise regular CME for the staff at NRH and also for the provincial clinics. A short overview of the project is included in 16.3 on page 166.

7.3. iPath in Developing Countries

Many of the applications described in the overview above include participants from developing countries. Over a third of all clinical cases presented on iPath originated from outside of Western Europe and US (c.f. table 8.1). Especially the diagnostic consultation groups for South East Asia (Cambodia, Laos, Bangladesh) and the Pacific as well as the Histopathology Forum with submissions from India, the Arab region, Armenia, etc. have been very active (c.f. detailed chapters below).

One of the objectives in the development of iPath was to ensure that medical specialists from different geographic regions, different time zones and especially with very different infrastructure can collaborate on the same set of data. A key functionality of iPath is that a consultation can be initiated by sending an email to the iPath-server, which is by far the most convenient way of data transfer in most low resource settings (c.f. table 8.1). For the consultants, it is often more convenient to work with the web interface of iPath. On the on-line site of iPath consultants can collaborate with each other and it is always visible what comments and diagnosis had already been given by others. However, all comments are sent back to the case submitter by email. These emails are generated automatically by iPath and they always include a full case report with the original case description plus all the comments given so far.

The concept of iPath makes it attractive for specialists from industrialized nations to use it beneficially for their own purpose. However, once they are familiar with iPath itself, it is often much easier to invite these specialists for collaboration with partners from developing countries.

7.4. Other Applications of iPath Software

While most telemedicine projects working with iPath are using the iPath-server at the University of Basel, there are some projects which are using the iPath software independently in their own IT environment.

One such installation was done at the Walter Sisulu University (WSU), the former University of Transkei (UNITRA), in Mthatha, South Africa, during my research stay there in 2003. Reasons to install a separate server at WSU were 1) increasing speed of access by eliminating the need for international bandwidth when using the server in Basel and 2) fostering regional organisation and facilitating the integration of telemedicine with the local health system.

This server is serving two main purposes. The first is a diagnostic tele-dermatology service which is available to all primary care facilities in the province. This service is now managed by the Telehealth Committee of the Department of Health of the Eastern Cape Province. Despite its availability the service was so far only used by three clinics with a majority of consultation from one general practitioner in Port St. Johns[22]. This is mainly due to the lack of digital cameras and computers with an Internet connection in the clinics and lacking training. The Department of Health has planned to equip approximately 10 clinics (personal communication by the director of the Telehealth Committee). However, the plans are still on paper only and have not been set into practice.

A second application of iPath at WSU is the field of problem-based learning for the students of the medical faculty. A major problem in the problem-based learning course has always been the availability of copies of all documents (x-rays, ultra-sound images and other illustrations) for all students. A selection of "learning problems" is now made available to the students through the iPath server.

Since other institutions can download iPath freely, we do not exactly know how often and for what purpose it is being used. However, the statistics on sourceforge[1], the website on which the source code is available, show that since the initial release in May 2001 there have been over 5'000 software downloads. Some of the institutions that use iPath include the Department of Pathology of the University of Pittsburgh Medical School (USA), the Argentinian society of cardiology, HealthNet Nepal, University of Udine (Italy) as well as several non-medical projects.

---
[1] http://www.sourceforge.net/projects/ipath

Usage Statistics ot the iPath Server at University of Basel

group name	members	cases	comments	comment per case	average case views	users seding cases	users seding comments
Brustzentrum Dresden	20	1375	20	0.01	15.49	10	6
Sihanouk Hospital Center of HOPE, Cambodia	66	1296	2861	2.21	31.17	7	35
South Pacific	79	499	1134	2.27	41.10	11	29
Oncology Center Lorrach	148	475	159	0.33	40.61	14	16
Bangladesh	65	346	1549	4.48	58.12	3	34
Feldstudie Telepathologie	22	325	390	1.20	16.07	12	17
Basel Atlas Daicker	5	247	532	2.15	30.27	1	4
Laos Vientiane	58	229	450	1.97	28.03	8	26
Histopathology Forum	242	225	1069	4.75	87.72	18	57
Basel Ophthalmologie	2	203	247	1.22	11.37	2	3
Telecytology Basel	16	168	700	4.17	54.29	1	8
Iran Pathology Group	43	163	671	4.12	45.99	5	24
Radiologie Lorrach	7	162	1	0.01	9.53	1	1
Ukrainian Swiss Perinatal Health Programme	113	159	308	1.94	60.53	27	40
AG Knochentumoren	52	140	993	7.09	118.66	28	29
AG Mamma- und Gynakopathologie der SGP	37	120	7	0.06	27.28	17	3
Expertgroup Pathology Basel	33	119	336	2.82	26.34	10	15
Knochentumor-Kommission	35	91	429	4.71	74.90	2	15
Dresden	15	83	168	2.02	24.72	2	9
WGS	16	78	28	0.36	30.41	12	4
Basel Ovarzentrum	3	70	0	0.00	16.27	1	0
Zurich Neuropathology	11	64	2	0.03	11.81	4	2
Ethiopia Pathology	29	61	141	2.31	47.92	2	13
Hematopathology	43	52	243	4.67	66.58	10	24
Qualitatszirkel BV Pathologen Sudbaden	6	51	4	0.08	7.65	2	1
Lithuanian Pathology online	50	46	116	2.52	55.89	9	10
PathoIndia	218	43	206	4.79	96.00	9	34
Pacific Pathology Group	91	41	126	3.07	43.34	7	15
Novartis Ophthalmology	3	41	8	0.20	12.41	2	1
Basel Surgical Pathology	84	40	13	0.33	34.55	3	8
Donetsk Regional Telemedicine Network	188	39	49	1.26	41.08	7	6
Nierenbiopsie-Diagnostik	27	36	4	0.11	31.19	2	1
Telecytology Honiara	19	35	143	4.09	52.51	4	10

group name	members	cases	comments	comment per case	average case views	users seding cases	users seding comments
Hepatopathology	8	35	16	0.46	12.94	4	4
Georgia Telemedicine	28	31	104	3.35	60.77	6	17
Basel Image Pool	31	28	5	0.18	53.54	3	4
SAKK 30/01 HOVON 43: Review	8	28	1	0.04	70.00	4	1
Umtata	32	27	85	3.15	61.52	2	15
Norodom Sihanouk Hospital, Cambodia	25	25	89	3.56	58.00	4	15
Basel Oberholzer	62	25	25	1.00	50.52	5	7
Romania Neonatology	46	23	47	2.04	55.57	9	15
Georgia Project	5	22	2	0.09	10.09	3	2
African Dermatology Forum	80	18	40	2.22	52.06	7	11
Basel Pathology Microscopy	46	18	14	0.78	44.67	5	7
RAFT - Forum	56	16	56	3.50	63.19	9	12
International Gynaecology, Pediatric and Adolescent Gynaecology	147	16	13	0.81	45.94	5	5
Leukemia Cytogenetics	8	16	23	1.44	31.31	1	5
Suisse DermoPratique	29	14	10	0.71	169.93	2	7
Basel Augenklinik	5	14	28	2.00	43.93	3	3
St.Gallen Pathologie	3	14	0	0.00	28.50	1	0
African Radiology and Ultrasound Forum	186	13	30	2.31	48.38	8	11
Swiss Lung Pathology Group (SLPG)	15	13	12	0.92	44.38	5	4
SAKK Leukemia group	19	13	17	1.31	31.38	4	4
SOL National Referral Medical Center (NRMC)	31	12	49	4.08	52.50	5	13
Vietnam	12	12	29	2.42	32.50	1	4
Zytologie UKE Hamburg	5	12	16	1.33	14.67	1	2
SOL National Health Education Center (NHEC)	26	12	1	0.08	19.17	2	1
Lung consult	16	11	13	1.18	90.00	3	5
Tropical Dermatology and Venerology Forum	23	11	19	1.73	35.82	2	4
Bern Leukoplakie-Projekt	6	11	8	0.73	49.91	3	3
International Perinatal Group	81	11	9	0.82	20.82	2	2
Port St Johns, ZA	3	11	5	0.45	23.09	2	1
Pathologie Thun	33	11	2	0.18	55.18	1	1
Suisse DermoTeach 2006	5	11	0	0.00	32.82	2	0

8. iPath – a Telemedicine Platform to Support Health Providers in Low Resource Settings

K Brauchli (1), D O'Mahony (2), L Banach (3) and M Oberholzer (1)

1) Department of Pathology, University Hospital Basel, Switzerland
2) Family Practitioner, Port St Johns, South Africa
3) Telemedicine Unit, University of Transkei, South Africa

This article has been published in *Stud Health Technol Inform* 2005;114:11-7.[22]

Abstract

In many developing countries there is an acute shortage of medical specialists. The specialists and services that are available are usually concentrated in cities and health workers in rural health care, who serve most of the population, are isolated from specialist support[50]. Besides, the few remaining specialist are often isolated from colleagues. With the recent development in information and communication technologies, new option for telemedicine and generally for sharing knowledge at a distance are becoming increasingly accessible to health workers also in developing countries. Since 2001 the Department of Pathology in Basel, Switzerland is operating an Internet based telemedicine platform to assist health workers in developing countries. Over 1800 consultation have been performed since. This paper will give an introduction to iPath - the telemedicine platform developed for this project - and analyse two case studies: a teledermatology project from South Africa and a telepathology project from Solomon Islands.

8.1. Introduction

Health providers like doctors and hospitals in developing countries often suffer from limited or non-existing access to specialists[18, 50, 75, 130]. For example, the National Referral Hospital (NRH) in Honiara, the only major hospital on Solomon Islands serves a population of approximately 450'000 people and there is not a single pathologists or dermatologist. In 2001, a simple histology laboratory was set up in Honiara. Microscopic slides are prepared in the lab and subsequently photographed with a digital camera and submitted via email to an Internet-based telemedicine platform located at the University of Basel, Switzerland. Several pathologists in Europe review these images and within 8.5 hours (median) a diagnosis is made available to the surgeon in Honiara[18].

Following the successful example of telepathology in Honiara, other projects started using that telemedicine platform and now there are approximately 70 consultations from developing countries every month. While pathology had been the first applications, there are now several teledermatology projects in Africa using this platform and also one large project for neonatology consultations in Ukraine.

In all these examples, telemedicine is not used directly by the patient but primarily by doctors and nurses who need the additional input from specialists to improve the services that they are delivering.

8.2. iPath - a Hybrid Web and Email Based Telemedicine Platform

Since 2001, the Department of Pathology of the University Hospital Basel has been developing the iPath software (http://ipath.ch), an open source framework for building web and email based telemedicine application[16, 21]. iPath provides the functionality to store medical cases with attached images and other documents into closed user groups (c.f. Fig 8.1). Within these groups, users can review cases, and write comments and diagnosis. Additionally, users can subscribe for notifications so that they get an automatic email if e.g. a new comment was added to one of their cases or if a new case is entered into a group.

Technically, iPath is a web application written in PHP. From the functionality it is somewhere between a content management system (CMS) and a group-ware tool. All users are organised into several discussion groups. Every discussion group has at least one moderator who can assign other users to the group and who can delete erroneous data. Thus, the system does not need to be administrated centrally as every group is administrating itself[18].

A very useful function of iPath, especially for areas with limited resources is the automatic email import. Users must once specify a group into which they would like to store cases sent by email. Then they can send a case to iPath as an ordinary email from any email client, typing the case title as the subject of the email, the clinical description as main text and simply attaching images.

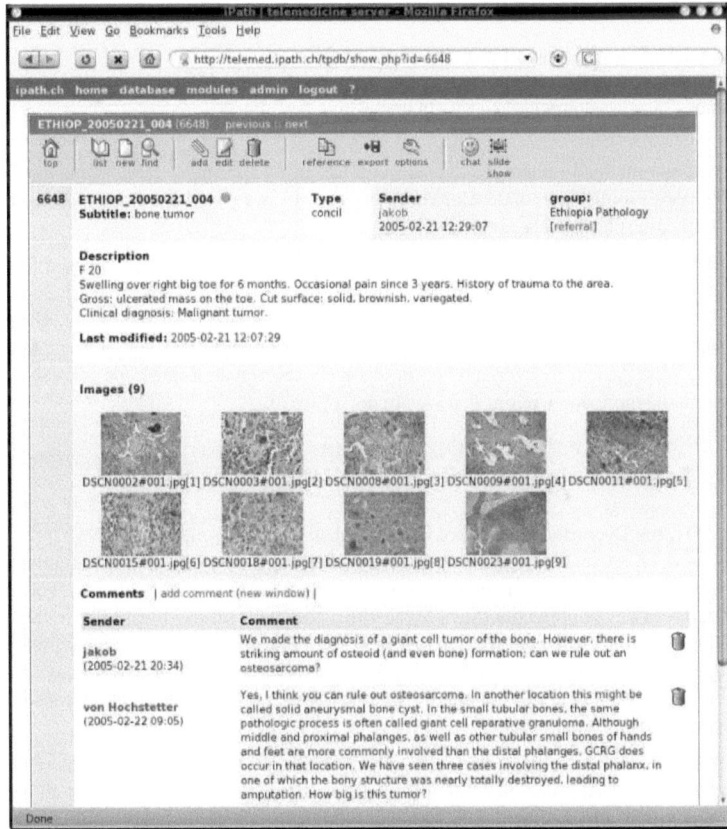

Figure 8.1.: A typical case in iPath. This is an example of a telepathology consultation from Ethiopia. At the top there is the general case information (sender, submission date) followed by a clinical description and an image gallery. Below, specialists can state their comments and diagnosis.

	Users	Cases	Images	daily logins (2004)	submissions by email
total	1213	5016	33247*	38	32.12%
developing countries	84**	1798	14006		74.17%

Table 8.1.: Usage statistics of iPath (24.12.2004)
* average file size 93KB. Besides images there were another 5864 files (pdf, powerpoint etc)
** only 47% of users specified country of origin.

iPath will automatically import such cases into the group specified. Table 1 illustrates that out of 1798 cases submitted from developing countries, 74% were submitted by email (compared to 32% of all case submissions world wide)

The iPath software has been released as an open source project that can be used for regional networks and by other projects. Currently, the main usage of iPath is the telepathology network at the University of Basel with over 1000 users world wide (c.f. Section 8.2.1). However, we are aware of iPath being used for regional telemedicine networks in South Africa, Nepal, North West US, West Africa, Switzerland and in Germany. However, as the code is freely available, there might be more applications that we are not aware of.

8.2.1. Telemedicine Platform at the University of Basel

Since 2001, the Department of Pathology of the University Hospital Basel, Switzerland, is operating an open telemedicine platform based on iPath[1] [16, 21, 18]. In the beginning the platform was mainly used for telepathology projects in Switzerland and for collaboration with some pathologists in developing countries. Meanwhile, the platform has over 1300 users and more than 5000 cases have been discussed so far (c.f. Tab. 8.1). Besides the pathology projects at our department, the platform is used for a wide range of application - from telepathology on Solomon Islands[18] to neonatology discussion in Ukraine (59 users) to teledermatology consultations in Africa (over 50 consultations)

Table 8.1 shows the basic usage statistics of this platform. By the end of 2004 there were 1213 users of which 84 had specified coming from a developing country (only 47% of all users specified a country of origin, so probably there are more form developing countries). Since the start of the project in September 2001 a total of 5016 cases with totally 33247 images have been sent to the server - on average 6.7 images per cases. The average image size was 93KB. If we look at developing countries only, there were 1798 cases submitted with a total of 14006 images - on average 7.7 images per case. For the year 2004 there was an average of 67 consultations from developing countries submitted every month. Figure 8.2 illustrates the origin of all these consultations. The largest contribution was from a telepathology project at the Sihanouk Center of Hope in Phnom Penh, Cambodia, which submitted over 700 cases.

---
[1] http://telepath.patho.unibas.ch

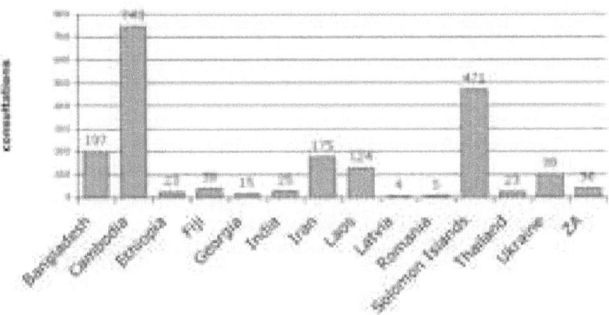

Figure 8.2.: Consultations submitted from developing countries since the start of the iPath server in Basel in September 2001. Two major parts of the submissions are from the telepathology projects in Cambodia (743) and Solmon Islands (471).

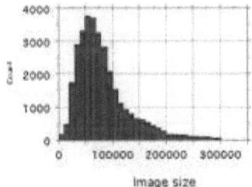

Figure 8.3.: Distribution of image size for consultations submitted to the Basel telepathology server. From September 2001 to December 2004 a total amount of 33247 images with an average file size of 93KB have been submitted (c.f. Table 8.1).

73

8.3. Case Studies

iPath is used for a wide range of telemedicine applications. To illustrate the practical application and outcome in low resource settings we will study two examples.

8.3.1. Teledermatology in Port St. Johns, South Africa

Port St. Johns is a small provincial town on the east coast of South Africa. It is located in the former Transkei which used to to be an 'independent' homeland during the apartheid. Now, the region is one of the poorest in South Africa. In the rural Port St Johns district, the population numbers about 75 000, the majority of which lives below the poverty line. Primary health care is provided mainly by nurses at state funded clinics, supported by general practitioners in the public and private sectors. In the last decade, the number of doctors in the district has varied between two and six. The referral hospital at Umtata is 100 Km distant but since 1998, has no had a specialist dermatologist permanently. At times the closest dermatology specialists was in East London, 350km from Port St. Johns. Hence, family practitioners have to diagnose and treat practically all dermatology problems (~15% of all consultations). To improve access for patients to dermatological care and to improve family practitioner clinical skills, a teledermatology project was initiated in April 1999[116]. The project started with email based store-and-forward teledermatology, and since 2002 it is using iPath. In the first year the server in Basel was used but since 2003 the Telemedicine Unit[116, 138] of the University of Transkei (UNITRA) in Umtata is running a regional telemedicine network based on the iPath software (http://telemed.utr.ac.za) which is now being used by the teledermatology project in Port St. Johns.

For the telemedical consultations images are captured with a digital camera (first an Olympus C-1400XL and later an Fuji 2 mega-pixel). Images were resized using Adobe Photoshop or GIMP[2]. In the beginning images were submitted by plain email with attached pictures. However, text and pictures easily got separated and misfiled. Thus patient information and images were compiled into an HTML page which worked well but was a very time consuming process. Finally, using iPath, clinical information and images are sent by plain email to the iPath server, where they are automatically inserted into a database and presented to the dermatology specialist in form of a concise web page. Besides the ease of use, the automatic email notifications system of iPath has also helped to reduce turnaround times. From an average response time of over 30 days, it is now at 6.5 days since consultation are done using the iPath platform at UNITRA.

Since 1999, 110 patients from Port St. Johns have been diagnosed using teledermatology. 76 patients where female and 34 male with an average age of 32 years. In 105 cases a telemedical diagnosis was possible and in 104 cases this assistance was judged helpful by the general practitioner (GP). For 57 cases, the telemedical diagnosis enabled an improvement of the treatment (unpublished data, an evaluation of the project is in preparation). The major outcome however is not only the direct improvement for the patient but also the fact that teledermatology helped

---

[2]Open Source image manipulation program - http://www.gimp.org

the GP to improve his skills in diagnosing and treating dermatology problems appropriately, or, citing the GP: "The number of cases dropped off over the years. This is definitely due to my improved skill in diagnosis due to learning."

8.3.2. Telepathology on Solomon Islands

The National Referral Hospital (NRH) in Honiara is the only major hospital in Solomon Islands, an independent state with approximately 450'000 inhabitants, tucked away in the south west of the pacific ocean. The NRH is the only referral hospital for the 8 provincial hospitals. The country has about 40 doctors but not a single pathologist and consequently tissue samples for histological examination have to be sent by airmail to the nearest pathology service in Brisbane, Australia and it is not unlikely that the doctors at the NRH have to wait 3-6 weeks before the histological diagnosis is returned from Brisbane. Besides, the state of Solomon Islands consists of over 900 islands, spread out over hundreds of kilometers. Patients from remote islands have to travel by boat for days to reach the hospital on the main island. For many patients it is difficult to return home to wait until a diagnostic result has arrived at the NRH and as a consequence, treatment decisions often have to be made without a histological diagnosis.

A small histology laboratory was established at the National Referral Hospital (NRH) in Honiara, Solomon Islands, in September 2001, allowing the preparation of H&E stained sections. Gross specimen are prepared by the surgeon, processed in the laboratory and the slides are usually ready two or three days later. From the microscopic sections prepared in this laboratory, digital photographs are taken using a Nikon CoolPix 990 Camera mounted on a Nikon OptiPhot 2 microscope. These pictures are usually scaled to approximately 600x400 pixels (typically 20KB - 70KB) then sent via email to the telepathology server at University of Basel[18].

During a two year period from January 2002 and December 2003 a total of 333 pathology consultations where submitted from NRH to the telepathology server in Basel. These consultations were submitted by email with a short clinical description and with images as attachments (average 8.8 images per consultation). In 50% of all consultations a first report from a pathologist was issued in 12h or less (cf. Table 8.2).

A major improvement in the project was the introduction of a virtual institute[21, 18]. A virtual institute is a group of experts with a duty plan. Every week one specialist is 'on call' and the iPath system automatically notifies the 'expert on call' about any new cases and also about new comments from other experts. Besides, the expert on call was asked to mark a diagnosis as final if in his or her opinion, a diagnostically conclusive response was possible based on the submitted material. This organisation helped to reduce the turn around time for diagnosis from 28h in the beginning (phase I in table 9.1) to 8.5h after the introduction of the virtual institute (phase II).

	Phase I	Phase II	total
Number of consultation	73	260	333
First response after (median)	28h	8.5h	12h
Consultation possible	93.2%	94.2%	94.0%
Additional images requested	24.7%	10%	13.2%

Table 8.2.: Telepathology consultations from National Referral Hospital in Honiara, Solomon Islands. Phase I are the consultations before the introduction of the virtual institute (cf. text) which is the time from January 2002 to October 2002. Phase II describes the situation from November 2002 to December 2003 after the introduction of the virtual institute. The second line indicates the median time between submission of the case by email and the first response from a pathologist. (Figure from Brauchli et al. 2004)

8.4. Discussion

When iPath was developed it was not primarily intended for telemedicine in low resource settings, however, it turned out that an easy to use telemedicine solution which does not have high demand on bandwidth can be a very helpful tool in developing countries. The platform has been very well used by health professionals working in developing countries to consult with specialists from other parts of the world to overcome the professional isolation often present in remote hospitals and to improve their skill and services they can deliver to their patients.

Looking at the usage of iPath over the past 3 years we can observe a number of different types of applications. Firstly there are remote consultations where typically a doctor at a remote hospital consults with a group of distant specialists. Secondly there is a growing number of general discussion groups (not only on iPath) where specialists working in isolation are sharing knowledge and experience with distant colleagues. Besides, iPath is more and more used for decentralised studies, where a number of partners are jointly collecting data on a special topic (research, quality control, etc). Data can be text, images and also custom forms for capturing structured data. The advantage of an Internet based solution is that every partner can at any time review the whole collection and compile statistics.

As iPath is developed as an open source project and distributed under the General Public License (GPL[3]) its use is not restricted to the telemedicine server of the University of Basel. The open source license allows other projects to use iPath and adapt it to their needs. As telemedicine is primarily used by specialists in centrally located institutions, it bears the risk of inducing a digital divide within a developing country if the periphery of the health system is not involved in the development of the network[53]. Besides there are often cultural differences and language barriers that are difficult to address in large international projects. The open source nature of iPath allows such adaptations and it is easily possible to reproduce working regional solutions as free and open source software can be adapted and distributed.

---

[3]Free Software Foundation - http://www.fsf.org

Part III.

Case Studies

9. Telepathology on the Solomon Islands–Two Years' Experience with a Hybrid Web- and Email-based Telepathology System

K Brauchli (1), R Jagilly (2), H Oberli (2), K D Kunze (3), G Phillips (4), N Hurwitz (1) and M Oberholzer (1)

1) Department of Pathology, University of Basel, Switzerland
2) National Referral Hospital, Honiara, Solomon Islands
3) Department of Pathology, Technical University of Dresden, Germany
4) Royal Brisbane Hospital, Queensland, Australia

This article was published in the Journal of Telemedicine and Telecare 2004; 10 (Suppl. 1): S1:14-17 [18]

Abstract

The National Referral Hospital in Honiara, Solomon Islands, has used an Internet-based system in Switzerland for telepathology consultations since September 2001. Due to the limited bandwidth of Internet connections on the Solomon Islands, an email interface was developed that allows users in Honiara to submit cases and receive reports by email. At the other end, consultants can use a more sophisticated Web-based interface that allows discussion of cases among an expert panel. The result is a hybrid email- and Web-based telepathology system. Over two years, 333 consultations were performed, in which 94% of cases could be diagnosed by a remote pathologist. A computer-assisted 'virtual institute' of pathologists was established. This form of organization helped to reduce the median time from submission of the request to a report from 28 h to 8.5 h for a preliminary diagnosis and 13 h for a final report. A final report was possible in 77% of all submitted cases.

9.1. Introduction

The National Referral Hospital (NRH) in Honiara is the only major hospital in the Solomon Islands, an independent state with approximately 450,000 inhabitants, in the south-west of the Pacific Ocean. The NRH is the only referral hospital for the eight provincial hospitals. The country has about 40 doctors but not a single pathologist and consequently tissue samples for histological examination have to be sent by airmail to the nearest pathology service, which is in Brisbane, Australia. With the decline in tourism after the civil disorder in 1999, transport to the Solomon Islands has become even more limited. It is common for the doctors at the NRH to wait for three to six weeks before a histological diagnosis is available from Brisbane.

Patients from remote islands have to travel by boat for days to reach the NRH on the main island. For many patients it is difficult to return home to wait until a diagnostic result has arrived at the NRH and, as a consequence, treatment decisions often have to be made without a firm histological diagnosis. Recent advances in telecommunications and telemedicine suggest ways of overcoming such problems. There is growing evidence in the literature that telemedicine is a feasible tool, even for countries with less well developed telecommunications infrastructure[44, 49, 130, 126, 140, 148, 167]. However, most of the reports deal with teledermatology[126, 130] and teleradiology[49], and there is little published experience in the field of telepathology in developing countries (only one citation in PubMed[101]).

We have therefore employed iPath, a hybrid Web- and email-based telemedicine system developed at the University of Basel[16, 21, 113]. Basically, iPath is a collaborative platform that allows a group of specialists to discuss cases; these typically consist of a clinical description and attached images or other multimedia objects. A special feature of iPath is that it offers static as well as dynamic telepathology[32] and also several interfaces for access to data. A user can work via an email or a Web interface, but there is also the possibility of interactive remote control of a robotic microscope. iPath is available as free software[69]. In October 2001, when the project was started, the Solomon Islands telecommunications provider had a 128 kbit/s link to the Internet, which had to be shared by all Internet users in the country. Because of this limited bandwidth, only static telepathology was practicable.

9.2. Methods

A small histology laboratory was established at the NRH in Honiara in September 2001 that was able to prepare sections stained with haematoxylin and eosin. The processing of the specimens was done manually, because the repair and maintenance of any specialized automatic equipment are difficult. The gross specimens are prepared by the surgeon and the slides are usually ready two or three days later.

From the microscopic sections prepared in this laboratory, digital photographs are taken using a digital camera (CoolPix 990, Nikon) mounted on a microscope (OptiPhot 2, Nikon). These pictures are usually scaled to approximately 600x400 pixels (typically 20-70 kByte) then sent via email to the telepathology server at the University of Basel.

The telepathology server in Basel is based on iPath[69]. Originally, iPath was developed as a consultation platform that offered access through a Web browser. However, the experience in the Solomon Islands led to the development of email-based access. The server can automatically import cases from email. The email text is stored as the case description and the attached images are placed in an image gallery.

These cases are then reviewed by an international group of pathologists. These pathologists are organized as a 'virtual institute' (VIRIN [21]) using the 'expert group' facility of iPath. As in a real institute, there is always one pathologist on call. When a new case arrives, the pathologist on call is automatically notified by email. The pathologist will then use the Web interface to review the case (Fig 9.1).

If a diagnosis can be given easily, the expert on call will simply write the diagnosis and label it as final. The system will then close the case and send the diagnosis automatically to the NRH by email. If the case is more complicated, the expert on call may state a preliminary diagnosis and then link the case to the VIRIN. Other members of the VIRIN are informed of the case by email and can report their opinion. These opinions are collected inside the VIRIN and are not directly accessible to the sender of the case. Finally, the expert on call will summarize the opinions of his or her colleagues and will write it down in the original case report. The referring doctors can read this diagnosis online or, in places where online Web access is difficult, the server can automatically send the final diagnosis by email.

9.3. Results

Between January 2002 and December 2003, 333 pathology consultations were submitted, by email, from the NRH to the telepathology server in Basel. These consultations comprised a short clinical description and images as attachments (an average of 8.8 images per consultation). In 50% of all consultations, a first report from a pathologist was issued in 12 h or less (Table 9.1).

The cases were submitted in two phases: phase I included all cases that were submitted before the introduction of the VIRIN in October 2002, while phase II included all cases submitted thereafter. During phase I, 73 cases were submitted. During this 10-month period, the pathologists were not organized in any particular way. Every pathologist would log into the system now and then and report on new cases. As Table 9.1 illustrates, in 50% of the cases a response from a pathologist was made no later than 28 h after submission of the case (on average within 32 h). In 25% of all submitted cases, the pathologists asked for additional images and requested a specific location and magnification for these images. Overall, in 93% of the cases, the submitted material was suitable for at least some degree of diagnostic interpretation.

One of the major problems with this method of collaboration was that the doctors in Honiara were left to surmise a conclusive diagnosis from the comments of the different pathologists.

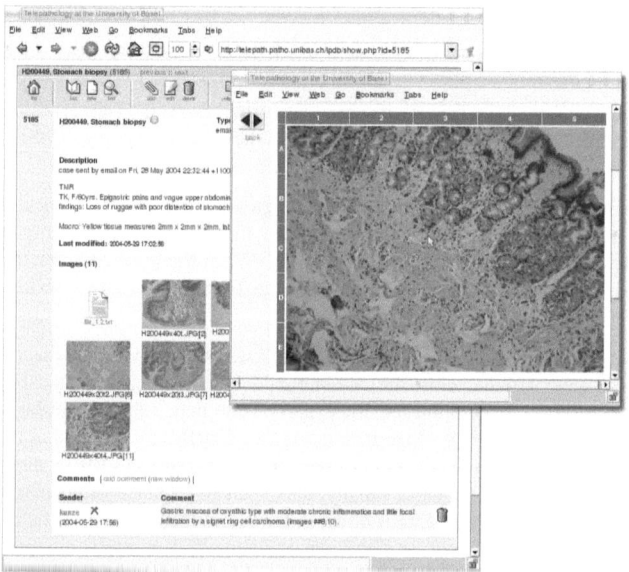

Figure 9.1.: The iPath Web interface. Every case consists of a header with information about the sender, date and title, followed by a clinical description and an image gallery. Images can be enlarged and the experts can enter their comments and diagnosis at the bottom of the page.

Table 9.1.: Telepathology consultations from the National Referral Hospital in Honiara

	Phase I[a]	Phase II[b]	Total
Number of consultation	73	260	333
Median time to first response (h)	28	8.5	12
Consultation possible	93.2%	94.2%	94.0%
Additional images requested	24.7%	10%	13.2%

[a]Phase I consultations took place from January 2002 to October 2002, before the introduction of the "virtual institute" (VIRIN).
[b]Phase II consultations took place from November 2002 to December 2003, after the establishment of the VIRIN.

Table 9.2.: Results of the virtual institute

	Results from 260 consultations (phase II)
Median (mean) time to final diagnosis (h)	13 (31)
Consultations with final report	77%
Second-opinion consultation in the VIRIN	17%
Median (mean) time to final diagnosis after VIRIN consultation (h)	74 (89)
Mean number of second opinions in the VIRIN	3.7

This led to the idea of forming a VIRIN, where second-opinion consultations were gathered in a closed discussion among the pathologists. Eventually one pathologist summarized the discussion and attached a conclusive response to the original case. This response was then automatically emailed to the doctors in Honiara.

The software to support the VIRIN was developed during October 2002 and in November the eight participating pathologists were reorganized as a VIRIN. A duty plan was prepared and each week one pathologist was on call. The iPath system automatically notified the pathologist on call about any new cases and also about new comments from other pathologists. In addition, the pathologist on call was asked to mark a diagnosis as final if, in his or her opinion, a diagnostically conclusive response was possible based on the submitted material.

In phase II, from November 2002 to December 2003, a total of 260 cases were submitted. In 50% of the cases the response time for a preliminary diagnosis was less than 8.5 h (mean 22 h) (Table 9.1). In 77% of all submitted cases, the pathologist on call submitted a final diagnosis (Table 9.2). The median response time for a final diagnosis was 13 h (mean 31 h). Eighty-three per cent of these cases were signed out directly by the pathologist on duty without further consultations, but in 17% a second opinion was requested from the VIRIN. On average, these cases received 3.7 comments from the VIRIN and for the cases discussed in the VIRIN a final diagnosis was available after a median of 74h (mean 89h).

It is noteworthy that, for the 260 phase II cases, a consultation was possible in 94% and only 77% were signed out with a final diagnosis. In other words, in 6% of the submitted cases the material was not sufficient for any kind of medical interpretation. The main reasons were technical problems or communication failures. For a further 17% of all cases, a preliminary medical interpretation was possible, but the material submitted did not suffice for the experts to reach a conclusive diagnosis.

9.4. Discussion

Telepathology dramatically reduces the time from specimen collection to results. The system established in the Solomon Islands is fast, convenient and cheap. The relatively quick results are a great relief for the patients, and for the relatives who are responsible for providing food and basic services for the patients while they are in hospital. The rapid results are also very helpful for the doctors and help to overcome the professional isolation which is a problem in remote places like the Solomon Islands. In particular, the direct interaction with the remote pathologist is a great benefit for the surgeons in Honiara. Finally, any reduction in hospitalization time should reduce costs and pressure on bed space.

The two years of using the system have shown several advantages of the hybrid system:

1. Consultants mainly work with the Web-interface and thus they can see all the cases and comments, and can easily identify difficult cases, such as those that have been erroneously submitted twice. Probably the most important advantage is that the experts can collaborate easily and discuss difficult cases within the expert group.

2. The email interface has proved to be very efficient in terms of both time and resources for the submission of cases and receipt of reports. The email interface does not implement all functions, but there is always the possibility of looking up all previous consultations using the Web interface.

3. System administration is very simple. Most settings can be adjusted by the users themselves.

There are also some disadvantages and limitations. Some training of the consultants is necessary for their proper collaboration in a virtual institute. The time necessary to organize and train the experts should not be underestimated. There are also some limitations that are inherent in all types of store-and-forward telepathology. The main problem is that it is possible for the operator in Honiara to miss areas important to the pathologist when taking pictures from the slides. This could be a pitfall, although a comparison (unpublished) of the telepathology diagnosis with the diagnosis based on reviewing the original glass slides has shown that in our series this is not a serious problem in practice.

In addition, taking pictures, processing and sending them require some time and therefore dedication. Thus it is important that the benefits are clearly visible in Honiara. Another specific limitation lies in the remoteness of the Solomon Islandsâit is much more difficult to get broken equipment repaired than in Europe. It is therefore important to choose equipment for robustness rather than performance.

There are also some areas that need to be improved:

1. A major limitation is the insufficient laboratory space that is available in Honiara. However, now that the positive results of the project have become obvious, it will be much easier to convince the hospital administration of the importance of such a laboratory.

2. A substantial number of samples are still sent to the pathology laboratory in Brisbane and it would be helpful to improve that collaboration. For reasons of quality control and ongoing training, collaboration with a relatively nearby pathology institute remains desirable.

3. There is a major need for cytology as well as histology services. It would be desirable to develop sampling procedures that allow an acceptable level of telecytology quality control for cytological diagnosis without a resident specialist.

4. A fully automatic scheduling system needs to be developed for the telepathology software (iPath). This should include adjustable, automatically supervised time limits for each sub-process (first response, final diagnosis) so that if an expert does not respond, another expert or an administrator is automatically informed. Such supervision would prevent some cases from being overlooked.

Our experience is that it is not difficult to produce goodquality slides in a simple histology laboratory and send them by email to an expert on the other side of the world to provide a diagnosis. Once set up properly, this is cheap and reliable, and would be useful for other remote places where there is no histopathology service.

Acknowledgements:

We thank Ana and Mike of the laboratory team at the NRH for their dedication and production of excellent quality slides. We also acknowledge the "Verein Meidzin im Südpazifik" and the Stanley Thomas Johnson Foundation for financial support of the project, and the University Computing Center (URZ) for technical support.

10. Diagnostic Accuracy of Telepathology on Solomon Islands

K Brauchli (1), R Jagilly (2), H Oberli (2), K D Kunze (3), G Phillips (4), N Hurwitz (1) and M Oberholzer (1)

1) Department of Pathology, University of Basel, Switzerland
2) National Referral Hospital, Honiara, Solomon Islands
3) Department of Pathology, Technical University of Dresden, Germany
4) Royal Brisbane Hospital, Queensland, Australia

Working Paper. Previewed for publication in the Journal of Pathology.

10.1. Introduction

Telepathology or tele-medicine at large is the practise of delivering medical services at a distance using telecommunication and information technologies (ICTs), thus transporting a specialists knowledge to the patient rather than the patient to the specialist. In telepathology such a service consist of delivering an anatomical pathological diagnosis based on microscopic images. This involves preparation of tissue and capturing of microcopic images at the remote site and the viewing and reporting of these images on a computer screen at the pathologists site. This can be achieved by means of dynamic telepathology[42, 111, 114, 156] using a remote controlled robotic microscope allowing the remote pathologist to select the field of view and magnifications or by the so called static-image or store-and-forward telepathology[107, 153]. More recently, virtual slide technology has become increasingly available, allowing the digitisation of a complete microscopic slide[58, 157, 158].

While dynamic telepathology and virtual slides are relatively expensive technologies and require a relatively high bandwidth, static-image telepathology has found broad application[7, 32, 38,

162]. Many studies on the diagnostic accuracy of static image telepatholoy have been performed and demonstrated the feasibility of the method[8, 39, 64, 82, 96, 97, 163] usually under the precondition that the submitting partner is himself a pathologist who has the necessary training of preparing tissue specimen and selecting relevant images under the micrscope. In the presented study we will look look at the situation where the submitting partner (non-expert) is a small hospital in a developing country with no resident pathologist.

Solomons Islands is an independent state in the south west-pacific of the ocean with a population of about 450'000 people living on the almost 1000 islands, forming a scattered archipelago of mountainous volcanoes and low-lying coral atolls stretching out over about 900 miles in the south west of Papua New Guinnea. The only major hospital in the country is the National Referral Hospital (NRH) in Honiara on the central island of Guadalcanal, staffed by about 20-30 doctors. There is one radiologist, but no pathologist, ophthalmologist or dermatologist. Consequently, tissue samples for histological examination have to be sent by airmail to the nearest pathology service, which is in Brisbane, Australia. With the decline in tourism after the period of civil disorder in 1999, transport to the Solomon Islands has become even more limited. It is common for the doctors at the NRH to wait for 3-6 weeks before a histological diagnosis from Brisbane is available.

In September 2001, a small histology lab was established at NRH and since then over 400 cases have been diagnosed over a telepathology link with the University of Basel, Sweitzerland. Usage of electronic communication and an efficient organisation of a tele-consultation service resulted in an average turn-around time for a diagnosis of approximately 12 hours after submission of the images[18, 22].

To establish the diagnostic accuracy of static-image telepathology under these conditions, the original glass slides were reviewed by an experienced pathologist and the results compared with the original telepathology diagnosis and where available with the report from the pathology department in Brisbane.

10.2. Material & Methods

10.2.1. Telepathology System

The telepathology link between the National Referral Hospital in Honiara (NRH) and the external pathologists is implemented using the iPath telemedine platform[18, 20]. As there is no resident pathologist at NRH, tissue specimen are selected and prepared in Honiara by the surgeon. In a small histology laboratory, all samples are processed to paraffin blocks, cut into sections of $4 - 6\mu m$ and finally stained using hematoxylin and eosin (HE (Hematoxylin Eosin)) by a team of two laboratory technicians.

Susequently, important areas of the slides are identified by the surgeon and photographed using a digital camera (CoolPix 990, Nikon) mounted on a microscope (OptiPhot 2, Nikon). These pictures are scaled to approximately 640x480 pixels (typically 20-70 kByte) then sent via email to the telepathology server at the University of Basel. Email transmission was done over a 33kb/s modem connection using a conventional telephone line and Solomon Telkom dial-up internet services. Although modem speeds of up to 56kb/s are available in Honiara, using the slower protocol at 33kb/s gave a more stable conenction and resulted in a better overall working experience for the surgeons in Honiara.

The iPath server imports these emails automatically every 15 min and transfers them into the case archive of the "south pacific" group which is one of about 100 working groups on the iPath server at University of Basel. A confirmation of the receipt is sent to NRH by email and the system administrator and the "pathologist on duty" are alerted by email. The expert pathologists are organized as a 'virtual institute' (VIRIN[21]) using the 'expert group' facility of iPath. As in a real institute, there is always one pathologist on call. When a new case arrives, the pathologist on duty is automatically notified by email. The pathologist will then use the Web interface to review the case (Fig 9.1).

If a diagnosis can be given easily, the expert on call will write the diagnosis and label it as final. The system will then close the case and send the diagnosis automatically to the NRH by email. If the case is more complicated, the expert on call may state a preliminary diagnosis and then link the case to the VIRIN. Other members of the VIRIN are informed of the case by email and can report their opinion. These opinions are collected inside the VIRIN and are not directly accessible to the sender of the case. Finally, the expert on call will summarize the opinions of his or her colleagues and will write it down in the original case report, which is again automatically forwarded to Honiara by email. All participating pathologists as well as the doctors at NRH can also access the whole case archive online at any time.

10.2.2. Review

Out of totally 333 case submissions from the first two years of telepathology at NRH a collective of 280 cases have been selected for reviewing the diagnostic quality. This collective includes all clinical cases for which the original glass slides could be retrieved from Honiara. Test cases and cases submitted for instructional or educational purpose only were excluded. For some of the cases included in the collective, a tissue sample was also submitted via airmail to the Queensland pathology deparment in Brisbane. For 125 cases the report from Brisbane was available.

Out of the total collective of 280 a review was possible for 261 cases. 1 case was prepared in the lab but never submitted by Telepathology (TP). 18 cases were excluded - in 14 cases the TP diagnosis was deferred and in 4 cases a review diagnosis was not possible on the available material. Finally, 261 cases of the initial 280 cases selected (93.2%) could be used for comparing the performance and diagnostic accuracy of the telepathology set-up.

For all 261 cases, the original glass slides were reviewed using conventional light microscopy by experienced pathologists (N.H.) of the Department of Pathology in Basel. All diagnosis were graded as simple, medium or complex - indictating the complexity of rendering a diagnosis for the presumed type of dissease based on H&E stained slides only. Additionally the reviewer classified every slide according to the quality of histologic preparation, sampling of tissue and selection of images.

All review diagnosis were then compared with the original TP diagnosis and where available with the diagnosis from Brisbane. All diagnoses were graded according to the discrepancy between the TP diagnosis, review diagnosis and diagnosis from Brisbane. The following grading was used:

1. identical (complete concordance)

2. minor discrepancy (indentical from clinical point of view)

3. moderate discrepacy (clinically not important with regards to treatment options available in Honiara)

4. marked discrepancy (clinically important)

Finally, the value of the telepathology diagnosis for the further treatment of the patient in Honiara was noted for all consultations.

10.3. Results

10.3.1. Material Submitted

The collective of 261 cases included in the review consisted of tissue samples form 137 female and 124 male patients with an age ranging from weeks old to 85 years (cf. fig.10.1). Tissue samples submitted represented all organ system (cf. table 10.1) and many typical classes of disease encountered in pathology (cf. table 10.2).

During the whole study 1844 images were submitted via TP – on average of 7.1 images per case ranging from 3 to 20 images per case. 1709 images (92.7%) were submitted with the original case submission. In 48 cases the pathologists asked for more images – 42 times for additional fields of view and 6 times for stronger magnification. In 20 of these cases additional images were submitted. In an addional 6 cases the surgeons from Honiara submitted additional images on their own account. A total of 135 images (7.3%) were submitted after the original case submission. Out of the total of 1844 submitted images, 1796 (97.8%) were histological photographs, 11 macro photographs of tissue samples, 22 photographs of the patient and 15 x-ray images.

In addition to the images submitted by the surgeons in Honiara, the pathologist also had the option of adding images to the case archive. In 6 cases the expert pathologist copied a low magnification image from Honiara and illustrated inside the regions from which additional images should be submitted.

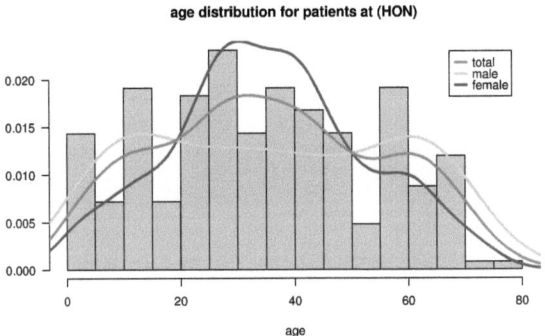

Figure 10.1.: Age distribution of the 261 patients included in the review study. The bars illustrate the effective frequency of each age group for the whole collective. The lines indicate estimated frequency distributions for the total collective and for male and female patients respectively.

Table 10.1.: Topography of submitted material submitted from NRH.

organ	cases (n=261)
Skin	36
Uterus (endometrium, cervix)	33
Breast	31
Soft tissue	28
Lymphnode	26
Prostate	14
Mucosa	11
Colon, Thyroid gland, Bone	10 (each)
Parotis (salivary gland)	6
Testes	5
Urinary Bladder	4
Ovary, Synovialis, Tube	3 (each)
D. deferens, Epidermis, Eye, Liver, Pharynx, Placenta, Tendon	2 (each)
aa Appendix, Bursa, Cartillage, Dura Mater, Ear, Esophagis, Gignivia, Lung, Mediastinum, Stomach, Tongue	1 (each)

Table 10.2.: This table summarises the diagnostic discrepancies identified with respect the the class of disease of the submitted material.

Class of disease		discrepancies			
	identical	minor	moderate	marked	total
Manlignat	49 (58%)	**6** (7%)	**13** (15%)	**16** (19%)	84
Suspicious for malig.	2	4	2	1	9
Benigne	18 (64%)	**5** (18%)	**3** (11%)	**2** (7%)	28
Others/undetermined	6	1	2	3	10
Inflammation	40	2	6	2	50
Scar, fibrosis	3		1		4
Hyperplasia	22		2	1	25
Tuberculosis	7				7
Inflammatory polyp	7		1		8
Abort	6	2			8
Endometrium (cycle)	6	1	1		8
Ulcer	3				3
total	**181**	**21**	**32**	**27**	**261**

10.3.2. Diagnostic Discrepancies

In 181 cases (69,1%) all diagnoses were identical (TP, review and Brisbane). 21 cases (8%) showed minor and 32 (12.2%) moderate discrepancies which are not clinically relevant under the circumstances given in Honiara. 27 cases (10.7%) exhibited a marked discrepancy between the TP diagnosis and the review diagnosis or the diagnosis from Brisbane. In 20 cases (7.7%) the TP diagnosis was markedly discrepant from the review diagnosis.

In 7 cases (2.7%) the review diagnosis and the diagnosis from Brisbane were different. These inconsistent diagnosis are mainly to be attributed to sampling problems. With small biopsies it is often a problem that the material could not be identically shared between TP and submission to Brisbane. In the case of fine needle aspiration (FNA (Fine Needle Aspiration)) for example, two different samples from the seemingly same tissue are submitted, one to Brisbane and one for TP. It is thus possible that only one contains tissue of a tumour. Such problems are not inherent to TP, but may occur in routine pathology.

A glance at the distribution of the discrepant diagnosis (c.f. tab.10.2) reveleals that most cases with marked discrepancies were encountered in the class of malignant tumours (16 out of a total of 27). In the class of malignant tumour 19% of the TP diagnoses show a marked discrepancy. For all other classes of disease together only 6.2% of all cases showed a marked discrepancy (total collective: 10.7%).

Based on the diagnostic and clinical problem, every case was graded as simple, medium or complex for diagnosis based on staic image telepathology. This grading quantifies the pathologosts "readyness" to commit a diagnosis based on the material and question presented on TP and was done indepenently form the review diagnosis. As figure 10.2 illustrates graphically, the majority

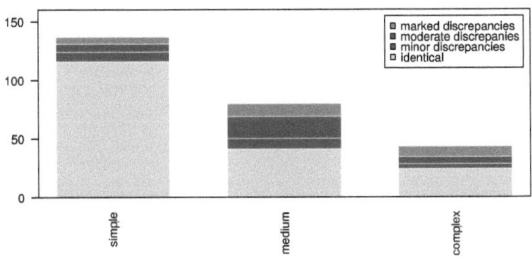

Figure 10.2.: The grade of the TP dianogsis. 53% were judged as simple, 31% as medium and 16% as complex. In the group of diagnosis graded as simple, the majority of TP diagnosis were identical to to the review diagnosis (84.7%) or showed only minor (5.8%) or moderate (5.1%) discrepancies. Only in 6 cases (4.4%) the TP diagnosis was markedly different from the review. For the medium and especially for the complex diagnosis the percentage for discrepancies was much higher - 13.8% for medium and 20.9% for the complex diagnosis.

of case that were regarded as simple the TP diagnosis was identical to the review diagnosis (85% complete concordance and only 4% marked discrepacies).

10.3.3. Problematic Cases

Table 10.3 illustrates the problematic cases that showed either a marked discrepancy or for which the value of the TP diagnosis was not graded as sufficient for treatment. Of the 42 cases considered as problematic 20 showed a marked discrepancy between the TP diagnosis and the review diagnosis based on the original slide.

The most commonly stated reason for discrepancy in table 10.3 is insufficient image selection which is mentioned 22 times. In 6 cases the reviewer came to the conclusion that no other reason than a mis-interpretation of the images in TP can be identified ("TP incorrect"). 2 cases exhibited insufficient histological quality, in 2 cases the tissue sample was not representative to address the clinical question of the submitting surgeon and in 2 cases the major problem was communication – in case 11 for example, the pathologists full answer was: "... The consensus diagnosis is Phylloides tumor. The dignity of this tumor should be assesed on multiple sections taken from multiple areas ...". The surgeon failed to submit additional images and the pathologists also neglected to specifically ask for additional images to be submitted.

10.3.4. Reasons for Diagnostic Discrepancies

In table 10.3 the reasons identified by the reviewing pathologist are listed. The most prominent reason is inadequate selection of images. "TP incorrect" indicates that no other explanation than

Table 10.3.: Problematic cases submitted from Honiara.

	discrepancy	Value of TP	class of disease	sex	age	TP	review	Reasons
1	marked	sufficient for treatment	malignant	m	19	Suspected metastatic germ cell tumor	Suspected metastatic seminoma	sampling error
2	marked	sufficient for treatment	normal	m	40	Vasitis nodosa	Normal vas deferens	
3	identical	insufficient for treatment	inflammation	f		Chronic active cervicitis	Chronic active cervicitis	
4	identical	insufficient for treatment	undetermined	m	1	Biopsy possibly not representative, no certain assessment possible	Biopsy possibly not representative, no certain assessment possible	
5	minor	insufficient for treatment	suspicious for mali	f	60	Tubulovillous adenoma with focal transformation into intramucosal carcinoma (G3)	High grade dysplasia (G3), areas highly suggestive of invasion	
6	minor	insufficient for treatment	suspicious for mali	f	26	Breast tissue. No malignancy. More samples	Breast tissue. No malignancy. More samples. Cystosarcoma phylloides cannot be ruled out	
7	moderate	insufficient for treatment	malignant	f	50	Suspicious for malignancy	Ductal carcinoma in situ	image selection
8	moderate	insufficient for treatment	inflammation	f	22	Delayed desquamation, hypersecretory endometrium	Chronic endometritis	communication
9	moderate	insufficient for treatment	inflammation	f	36	Premenstrual endometrium. No squamous mucosa	Chronic endometritis. No squamous mucosa	
10	moderate	insufficient for treatment	malignant	m	65	Pseudocarcinomatous hyperplasia. DD: Well differentiated squamous cell carcinoma	Squamous cell carcinoma	image selection
11	moderate	insufficient for treatment	malignant	f	14	Phylloides tumour of uncertain dignity	Malignant phylloides tumor	communication
12	moderate	insufficient for treatment	malignant	f	39	Suspected papillary carcinoma of the thyroid	Papillary carcinoma of the thyroid	image selection
13	moderate	insufficient for treatment	inflammation	f	25	Nonspecific chronic lymphadenitis	Lymphadenitis with microabscesses	image selection
14	moderate	insufficient for treatment	malignant	f	32	DD: Squamous cell carcinoma in phylloides tumor; Squamous cell carcinoma of the skin	Infiltrating ductal carcinoma of the breast	image selection
15	moderate	insufficient for treatment	inflammation	f	25	Proliferative endometrium	Chronic endometritis	image selection
16	moderate	insufficient for treatment	malignant	f	53	Suspicious for squamous cell carcinoma. No clearcut invasion	Squamous cell carcinoma	image selection
17	moderate	insufficient for treatment	suspicious for mali	m	38	Squamous cell carcinoma	Possibly squamous cell carcinoma, additional sections are needed to exclude keratoacanthoma	keratoacanthoma
18	moderate	insufficient for treatment	malignant	f	43	Highly suggestive of squamous cell carcinoma	Squamous cell carcinoma	
19	moderate	insufficient for treatment	malignant	f	44	Carcinoma of the breast	Lymphnode with metastasis of a breast carcinoma	failed morphological observation
20	marked	insufficient for treatment	malignant	f	50	Fibrous mastopathia. Focus of uncertain dignity	Carcinoma of the breast	image selection
21	marked	insufficient for treatment	inflammation	f	11	Normal appendix	Appendix with some evidence of inflammation. Severe periappendicitis	image selection
22	marked	insufficient for treatment	malignant	f	25	Normal skin	Well differentiated squamous cell carcinoma	image selection
23	marked	insufficient for treatment	suspicious for mali	m	50	Inflammatory polypous lesion. No dysplasia or neoplasia	Hyperplastic oral mucosa with mild to moderate dysplasia. Areas suggestive of invasion	image selection
24	identical	misleading	inflammation	m	15	Reactive changes. Fasciitis?	Reactive changes. Fasciitis? Biopsy probably not representative	
25	moderate	misleading	malignant	f	41	SCC: Squamous cell carcinoma	paget's disease and invasive carcinoma	image selection
26	moderate	misleading	malignant	m	30	Metastatic germ cell tumor	Metastatic poorly differentiated carcinoma	TP incorrect
27	moderate	misleading	scar, fibrose, granu	m	22	fibrous callus, no granulocytes	fibrous callus with acute and chronic infection, granulocytes	image selection
28	moderate	misleading	inflammation	m	22	granulation tissue with chronic inflammation, no granulocytes	soft tissue with chronic and acute inflammation, granulocytes	image selection
29	moderate	misleading	hyperplasia	f	40	In situ melanoma	Lentigo benigna. No malignancy	image selection
30	marked	misleading	benigne	m	12	Suggestive of a malignant tumor, probably malignant non Hodgkin lymphoma. More images needed	Pleomorphic adenoma	image selection
31	marked	misleading	malignant	m	49	Dermatofibrosarcoma protuberans	Squamous cell carcinoma	TP incorrect
32	marked	misleading	malignant	m	50	Lymphadenitis	Malignant non Hodgkin lymphoma, possibly T-cell angioimmunoblastic	TP incorrect
33	marked	misleading	malignant	m	65	Benign hyperplasia of the prostate	Carcinoma of the prostate	TP incorrect
34	marked	misleading	malignant	f	14	Tissue from breast. No malignancy	Fascia. No metastases	tissue sampling
35	marked	misleading	malignant	m	68	Soft tissue with heterotopic bone. No metastases	Metastases of a prostatic carcinoma	histology + image select
36	marked	misleading	undetermined	m	70	No malignancy	Prostata with atypical cells	histology
37	marked	misleading	malignant	m	38	No malignancy. Asking for additional images	Squamous cell carcinoma of the skin	image selection
38	marked	misleading	malignant	m	61	Moderate chronic prostatitis. No malignancy	Poorly differentiated carcinoma of the prostate	TP incorrect
39	marked	misleading	malignant	f	60	Condyloma accumminatum. No carcinoma	Condyloma accumminatum. Additionally squamous cell carcinoma	image selection
40	marked	misleading	malignant	m	50	No evidence of carcinoma	Poorly differentiated carcinoma of prostate	image selection
41	marked		malignant	f	43	No tumor. No malignancy	Cervical intraepithelial neoplasia grade 2 (CIN2)	image selection
42	marked		malignant	m	65	No evidence of malignancy	Carcinoma of the prostate	image selection

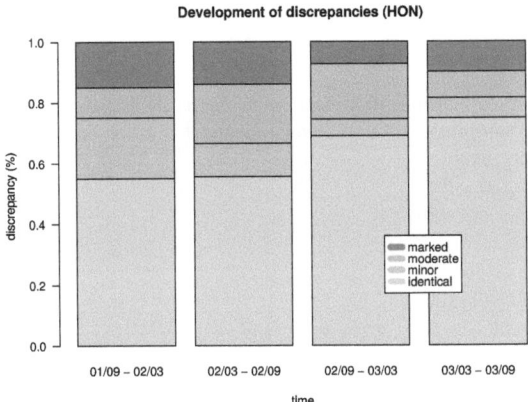

Figure 10.3.: Development of diagnostic discrepancies over time. The rate of consultation with identical diagnosis in TP and review increased from 55% in the first semester to 74.8% in the last semester. Concsultations with discrepand diagnosis dripped from 15% to 9.9%.

a failure on the side of the pathologist could be identified.

To analyse possible reasons for diagnostic discrepancies, the following four characteristics were evaluated for all 261 cases: *tissue sampling*, *quality of histology*, *image selection* and *communication*. Table 10.4 summarises the results of this analysis. To analyze the correlation between these factors and the diagnostic discrepancies, a multi-field χ^2 was performed on all tables. The quality of histology does not show any significant correlation at all. Tissue sampling scored a χ^2 of 26.5 which indicates a significant correlation (p<0.01). Communication problems and especially problems with image selection occur much more often in cases with marked discrepancies between TP and review – χ^2 test shows a highly significant correlation (p<0.001).

10.3.5. Development Over Time

The study includes case submission from Septmeber 2001 to September 2003. In order to look at the development of diagnosis over time, we grouped the consultations into 4 semesters (groups of 6 month each). Figure 10.3 illustrates the development of the diagnostic discrepancies over time. The number of consultations with identical diagnosis in TP and review increased from 55% in the first semester to 74.8% in the last semester. It should be noted here that the absolute amount of consultations also increased over time.

A similar improvement can be observed with regards to the quality of image selection by the surgeon in Honiara who had to select a number of relevant images for submission to the pathologists. The proportion of consultation for which the selection of images was rated "relevant" by the reviewing pathologist increased from 35% in the first semeseter to 68.7% in the last semester.

95

Table 10.4.: Reasons for discrepant diagnosis.

Tissue Sampling	total	identical	minor	moderate	marked
representative	175	133	14	16	12
representative for diagnosis, not sufficient samples	25	13	3	7	2
not representative	12	5	1	1	5
no statement/other	49	30	4	8	7

$\chi^2 = 26.5$ (significant, p<0.01)

Quality of Histology	total	identical	minor	moderate	marked
very good	107	74	12	13	8
good	93	67	5	11	10
moderate	52	34	4	8	6
insufficient	9	6	1		2

$\chi^2 = 4.29$ (not significant)

Image Selection	total	identical	minor	moderate	marked
all relevant changes documented	156	132	8	9	7
appropriate for diagnosis, some findings missing	32	22	5	3	2
moderate in quality or quantity, but sufficient for diagnosis	40	25	6	6	3
insufficient	33	2	3	14	14

$\chi^2 = 103.9$ (p<0.001)

Communication	total	identical	minor	moderate	marked
efficient	223	172	19	16	16
deficiency on case submitter's side	26	6 (1)	1 (1)	10 (10)	9 (8)
deficiency on both sides	12	3 (2)	1 (1)	6 (5)	2 (2)

(values in brackets indicate major communication problem)

$\chi^2 = 64.6$ (p<0.001)

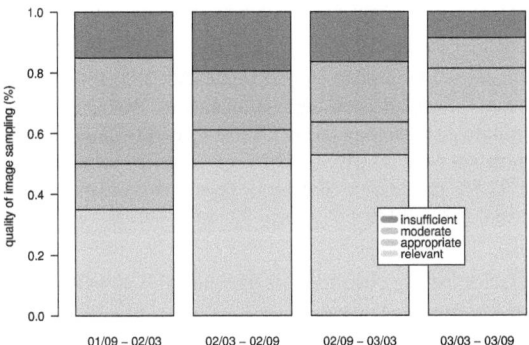

Figure 10.4.: Development of the quality if image selection over time. The proportion of consultation with insufficient selection of images was reduced from 15% in the first semester to 8.7% in the last semester. The amount of consultation for which the selection of images was rated as relevant almost doubled from 35% to 68.7%.

10.4. Discussion

Many hospitals in developing countries suffer from a lack of medical specialists. Telemedicine may provide a possibility to overcome this limitation through collaboration with external specialists. A major limitation of telemedicine is the fact that some task that are conventionally carried out by or under the supervision of one specialist need to be divided between different, geographically separated partners. In case of the telepathology diagnosis at NRH, preparation of the specimen (grossing) is done by a local surgeon with no special training in pathology. The processing of tissue and preparation of slides is done by the laboratory in Honiara and finally the selection of images under the microscope is carried out by the local surgeon. These tasks are conventionally carried out or supervised by a fully trained pathologist.

This study analyses the consequences of such a separation of tasks and its effect on the diagnostic accuracy of telpathology. Out of the 261 telepathology consultations studied in this paper, 20 cases (7.7%) showed a clinically relevant discordance between the TP diagnosis and the review diagnosis[1]. 181 cases (69.1%) showed complete concordance between TP and review and 21 cases (8%) had minor and a further 32 cases (12.2%) had moderate discrepancy.

The selection of images is generally considered as the most problematic step in static image telepathology[95, 150, 163]. This obersvation is also supported by this study, where we could

---
[1] If we also consider the diagnosis from the Queensland Health Pathology Department in Brisbane the figure is increased to 27 cases (10.3%). The difference between the diagnosis in Brisbane and our review are difficult to asses as they have been carried out under very different conditions and the material examinaed was not always completely identical because in some occasions the surgeons from Honiara had taken two biopsies and submitted one to Brisbane and prepared the other for telepathology.

demonstrate that the accuracy of a diagnosis correlates strongly with the quality of image selection. This is especially problematic if the case submitter who selects the images is not a trained pathologist.

These results are comparable to other similar studies[38, 123, 150, 163, 178]. In the probably largest study of static-image telepathology, published by Williams et al. in 2001[163] at the Armed Forces Institute of Pathology (AFIP), 1250 second opinion consultations from pathologists form all continents were reviewed. 73.7% of all reviewed submissions showed complete concordance and 97.3% recorded a clinically relevant concordance. A more recent study of telepathology in rural India[38] that evaluated 92 consultations, reported marked discrepancies in 9.7%.

Another limiting factor that we observed was the quality of communication. In some cases the pathologists asked for additional images or further clinical information, but the request was not followed by surgeon in Honiara. Similarly in some cases where additional images or clinical data were added by the surgeon the pathologists failed to revise their diagnosis. Ideally a telepathology tool should help the users to solve this kind of communication problem. Based on these experiences the iPath platform was extended with some tools that help to improve the flow of communication between pathologist and case submitter. A major step was the introduction of a "virtual institute"[18, 21] in which each case is assigned to an expert on duty who is then responsible for adequate processing.

A virtual institute (VIRIN) is a group of experts that may be responsible for a number of "case submission" groups. All cases that are submitted to iPath to any of the submitting groups are labeled as *new* (status is red) and are automatically placed on a review list for the expert on duty. The expert can either sign out the case or refer it to the other members of the expert group. If a diagnosis is possible on the material submitted, the expert on duty will add the diagnosis as "final comment" to the case and iPath will automatically close the case (status green) and remove it from the review list. If additional images or further clinical information are necessary, the expert will change the case status to *in review* (yellow) which indicates to both submitter and expert that there are pending requests. If the case submitter has an additional question for the pathologists he can also change the status of a previously closed case back to *in review*. All cases labelled as *new* or *in review* are automatically placed on the review list which is available for the expert on duty. Cases labelled as *closed* are removed from the review list.

If the case is difficult, the expert on duty has also the possibility to refer the case to the whole team of experts. On referral, iPath will automatically notify all experts in the group by email. The experts can then discuss the case within their own group and the expert on duty has the obligation to summarise this discussion and write a short report back on the original case which is then automatically sent by email to the case submitter. This procedure has proved to be very valuable not only for improving the communciation but also for reducing the turn-around time[18].

Interestingly, in this study the quality of histological preparation of the slides did not affect the diagnostic accuracy at all. This is probably mainly due to the fact that the laboratory in Honiara has generally produced high quality slides – despite the very limited infrastructure, 77% of the slides were of good or excellent quality and only 5 slides (2%) were of insufficent quality.

Additionally, it is worthwhile noting that the quality of the images was generally good, too. In contrast to the AFIP study[163], we observed almost no images that were not in focus. Selection of images and communication were the only problematic areas.

These observations are important due to the fact that the selection of images and especially the issue of communication can be addressed mainly by continuous traning. Throughout the project the pathologists tried to give some instructions on how to improve image selection where necessary. This effort is refelcted by the improvement of image selection as well as the reduction in diagnostic discrepancies over time (cf. fig.10.3 and fig.10.4).

10.4.1. Telepathology and Continuous Medical Education

Besides the immediate diagnostic results of these teleconsultations there is the aspect of continuous medical education and the problem of professional isolation that should be addressed here, too. Lack of specialists in a tertiary hospital in a developing country does not only affect the care of patients. Specialists play an important role in providing guidelines and continuous medical education for the practising doctors. Health professionals in a hospital without medical specialists have hardly the possibility of discussing a difficult case with an experienced specialist and thus there is often a feeling of professional isolation. Compared to submitting specimen by airmail, telepathology does not only reduce the turn-around time for diagnoses, but also offers the surgeons the possibility to interact with the pathologist.

Another important impact of the telepathology link was that the input of pathologists was very much needed to establish the histology laboratory in Honiara in the first place. Without a minimal histology lab it will be most difficult for the NRH to ever recruit a pathologist, but without the feedback from a pathologist there is no real incentive for the lab to produce good quality slides – a typical vicious circle. In Honiara, however, there is currently a young physician doing a specialisation in pathology at the University of Papua New Guinea and in Australia. One of his preconditions for coming back to Honiara was the availability of a minimal histology lab with technicians that are able to produce good quality histological slides. And possibly the now existing telepathology link that allows access to colleagues is only a first step in establishing a proper local pathology department.

10.4.2. Beyond (Tele-)pathology

The main objective of the telepathology project with the NRH was certainly to improve access to pathology diagnoses. However, the project had some interesting side effects. By using telepathology the doctors at NRH started to become familiar with telemedicine in general. Since iPath is an open telemedicine platform, which is not at all limited to pathology, the staff at NRH developed some ideas for extending the use of telemedicine (using iPath) to the provincial hospitals and clinics throughout Solomon Islands offering doctors and nurses the possibility to consult with the doctors at the NRH and some of their affiliated colleagues overseas, mainly in Australia.

Some hardware had already been available from other projects[106, 141] and through private initiatives. Some of these projects focused on linking rural clinics in developing countries directly with specialists in the UK and Australia[141, 148]. While this may work well initially, there are some inherent problems. Not all overseas specialists are familiar with the specific situation in a country like Solomon Islands and may not be readily able to give advice that is relevant under the local conditions. Besides, for complex cases that cannot be treated in the provincial clinic, the major question is usually if and how a patient can be referred to the next level hospital within the country. For this it is clearly necessary to consult with the doctors from that hospital.

The new project for a national Solomon Islands telemedicine network which is currently in a planning phase puts the NRH in the centre of the network. All consultations from provincial level are primarily directed to the NRH. However, it is planned to etsablish a network of external specialists for different medical fields so that the staff at the NRH can refer certain cases to the appropriate specialists. Technical support for this network will be provided through People First Network[2], a UNDP (United Nations Development Programme) project for sustainable rural networking on Solomon Islands.

It is also envisaged to utilize this network in collaboration with St. Vincents Hospital in Syndney, with which the NRH has an agreement to send a limited amount of patients for advanced treatment. The telemedicine network could greatly help to improve the decision making on which patients to send to St. Vincents. Finally, the network should also be available to organisations such as the Pacific Islands Project[3], which regularly sends medical specialists to some pacific countries for short term visits.

10.5. Conclusions

This study clearly demonstrates that static image telepathology is a feasible tool for remote diagnosis in clinical pathology for hospitals with limited resources. Even under restricted circumstances with no fully trained pathologists at the submitting hospital, it is possible to achieve a reasonable diagnostic accuracy. Selection of relevant images by the physicians or surgeons that are submitting the cases and the communication with the remote specialists have been identified the as the major limitations. However, both factors can be addressed with continuous traning which results in improved diagnostic accuracy and reduced turn-around times. The crucial aspects of successful telemedicine in developing countries are not technical limitations but training and continuous development of the people involved. Besides the diagnostic support, telemedicine is a most useful tool for overcoming professional isolation and for providing countinuous medical education in remote locations.

---

[2]http://www.peoplefirst.net.sb/
[3]http://www.surgeons.org/AM/Template.cfm?Section=Pacific_Islands_Project

Acknowledgement

The telepathology project at the NRH was financially supported by the "Verein Medizin im Südpazifik" in Brienz, Switzerland, and the Stanley Thomas Johnson Foundation in Bern, Switzerland.

11. Telepathology at the Sihanouk Center of Hope, Cambodia

K. Brauchli (1), C.S. Vathana (2), C.S. Ang (2), C. Häner (2), T. Kwakpaethoon (3), G. Stauch (4), K.D. Kunze (5), M. Oberholzer (1)

1) Department of Pathology, University of Basel, Switzerland
2) Sihanouk Hospital Centre of HOPE, Phnom Penh, Cambodia
3) Department of Pathology, Rajavithi Hospital in Bankok, Thailand
4) Institute of Pathology in Aurich, Germany
5) Department of Pathology, Technical Univeristy of Dresden, Germany

> Working Paper. Previewed for publication in the Journal of Pathology.

11.1. Introduction

With the rapid spread of information and communication technologies (ICTs) around the world and with increasing availability of Internet in most countries, telemedicine has been promoted as a promising tool to address deficiencies in delivering health care in developing countries[44, 53, 59, 167, 182], especially for accessing specialist advice and second opinion consultations in fields like dermatology, radiology or pathology, where there is an acute shortage of specialists in many developing countries.

Over the past 10 years, telepathology (TP) has become a well-established tool to deliver histological and cytological diagnoses at a distance and to access second-opinion consultations in many parts of the world[21, 35, 40, 41, 112, 163]. While many studies have demonstrated the feasibility and the the accuracy of diagnoses in developed countries[115, 124, 150, 163], there is little to no published evidence about the efficiency and diagnostic accuracy of telepathology consultations from developing countries. A major difference is that in many hospitals in developing countries that do require pathology services not have any staff trained in pathology at all, thus telepathology services must rely on surgeons or other physicians doing the basic pathology tasks of tissue sampling and image selection. Problematic areas for delivering diagnoses through TP without an on-site pathologist are the sampling of the original tissue, the laboratory process, selection of images and finally the communication between physician and remote pathologist.

As the provision of telepathology diagnosis without an on-site pathologist is a controversial issue[117], this study aims at evaluating the diagnostic accuracy of such a service by reviewing 212 telepathology diagnoses delivered to the Sihanouk Hospital Centre of Hope (SHCH) in Cambodia between January 2003 and January 2004. Each of the original diagnoses issued through the iPath telepathology server (http://telemed.ipath.ch) of the University of Basel was compared to an independent review diagnosis based on the original glass slides.

11.2. Material & Methods

The SHCH (Sihanouk Hospital Centre of Hope) is located in the Phnom Penh, the capital of Cambodia.There are 31 beds (7 in the emergency, 11 in the medical ward and 13 in the surgical ward). The staff consists of 25 medical doctors and 6 surgeons seeing around 200 to 250 patients a day with follow up, and 30 to 40 new patients. Besides providing health care the hospital is also committed to the further education of Cambodian doctors and health care professionals and is running many teaching activities.

Although the hospital is well staffed compared to other hospitals in Cambodia there is no fully trained pathologist. However, for clinical decisions on patients' treatment a pathologist's diagnosis and advice is often required. One of the physicians working in internal medicine (S.C.V.) had been trained in pathology in Belgium for one year in 1997. Today his position in the hospital is that of a lead physician and he is doing pathology via telepathology (TP) on two days a week. Besides telepathology, SHCH is also involved in other telemedicine activities[15].

11.2.1. Laboratory

In July 2002 a histology lab was set up at the Sihanouk Hospital with the help of colleagues from Aurich, Germany. The lab is equipped to prepare histological slides as well as cytological samples using conventional staining (H&E, PAP, Gysema, etc).

Images are selected by the physicians at SHCH and captured with an Olympus BH-2 Microscope equipped with a KLUG KamPro 04+ 1/2" SVHS CCD digital video camera. The camera is connected to a PC (Intel Celeron, 40GB HDD, 256MB RAM, Window XP Professional) using a FALCON single channel frame-grabber card [1]. The hospital had a permanent wireless Internet connection which was later upgraded to an ADSL-connection.

---

[1] IDS Imaging Development Systems GmbH, Obersulm, Germany. http://www.ids-imaging.de

11.2.2. Telepathology

Images are transmitted to the iPath server of the University of Basel (http://telemed.ipath.ch)[18, 20] normally by plain email. These emails are imported by the iPath server into a closed user group and the specialists are automatically alerted by email. The specialists can review the cases on-line on the iPath-server and enter their diagnosis using the web interface of the server. Diagnoses entered by the pathologists are automatically sent to SHCH by email.

Since its inception the TP lab at SHCH has submitted over 1300 consultations to the iPath server at the University of Basel. These consultations have included over 16'000 images and received over 2'900 diagnostic comments.

The pathologists are organised on a fixed duty plan. Every week, one member of the specialist group is "on duty". Whenever a new case is received by the iPath server, the expert on duty is automatically informed by email. This expert should then review the case within 24h. If the presented case and the clinical question are within the field of specialty of the expert on duty, he or she will write a diagnosis and mark it as "final" whereby the case is closed and the diagnosis is automatically transmitted to Cambodia by email. If the expert on duty does not feel competent in this particular case, he or she has the option to refer the case to the entire group of experts. When referring a case the expert on duty immediately writes a short preliminary report with a remark that the case is referred to the expert group. The other members of the expert group are then automatically notified by email and will discuss the case among themselves. After two to three days the expert on duty will summarise the comments by the fellow experts and will write a summary report on the original case which will again be automatically sent to Cambodia by email. This functionality of iPath is often referred to as "virtual institute"[18].

For this study 212 specimen submitted from January 2003 to January 2004 were reviewed. For all these cases the original glass slide was reviewed by a senior pathologist in Germany (KDK), and the diagnosis was recorded independently from the original TP diagnosis. Besides diagnosing all glass slides the cases were graded according to the diagnostic difficulty (simple, medium, complex). Additionally the sampling of tissue was also evaluated. After the review all diagnoses were compared to the original TP diagnosis and classified as completely identical or as diagnosis with minor, moderate or marked discrepancy.

In addition we reviewed all TP consultations and evaluated the selection of images by the non-expert and noted communication problems that occurred for some consultations.

11.3. Results

In the period of the study 212 specimen were submitted for Telepathology (TP) consultations. These consultations included specimen from 124 female and 88 male patients with an age ranging from 13 to 81 years (cf. fig.11.1). In total 2'703 clinical images were included with original case submissions. In 64 cases the remote pathologist asked for additional images. In 19 cases additional images were submitted (256 images or 8.8%). On average a case consisted of 14

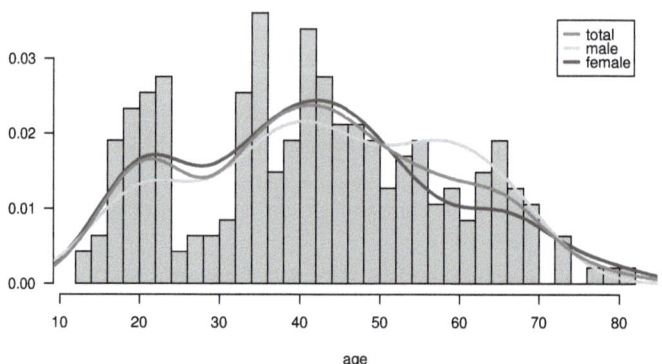

Figure 11.1.: Age distribution of the 212 patients included in the study. The bars illustrate the effective frequency of each age group while the lines indicate estimated frequency distributions for the total collective and for male and female patients respectively.

images. 2'914 (98.5%) histological photographs, 12 photographs of the patient and 33 x-ray images were submitted during the study.

149 cases were diagnosed within the same day they were submitted and another 40 cases were diagnosed the next day. In 53 cases the diagnosis required more than 2 days with a maximum of 10 days. The average response time for a TP diagnosis was 27h.

11.3.1. Diagnostic Discrepancies

For 179 specimen (84.4%) the TP diagnosis was completely identical with the review diagnosis on the original glass slide. 18 specimen (8.5%) showed minor discrepancies (clinically identical) and 5 specimen (2.4%) showed moderate discrepancies which were not clinically relevant. 7 cases (3.3%) exhibited a marked discrepancy (clinically relevant) between the TP diagnosis and the review diagnosis. 3 specimen were classified as "other".

Table 11.1 illustrates the distribution of discrepant diagnoses according to the class of diseases. The percentage of identical diagnoses was slightly higher for cases with a benign lesion than those with malignant lesions.

From the 212 submitted consultations a final diagnosis was deferred on TP in 12 occasions (6%). The reviewing pathologist rejected the material in 6 of these 12 cases as insufficient for diagnosis mainly due to tissue sampling not being representative to answer the clinical question. For the other 6 cases, deferred only on TP, the following reasons were given: twice insufficient image quality, once insufficient preparation, once a re-biopsy was recommended and twice no reason

Table 11.1.: Distribution of diagnostic discrepancies for different classes of disease. The diagnostic accuracy was slightly better for benign and other non-malignant lesions than for malignant lesions.

Class of disease	discrepancies				
	identical	minor	moderate	marked	total
Malignant	**68** (83%)	**10** (12%)	**1** (1%)	**3** (4%)	82
Suspicious for malig.	1				1
Benign	**24** (89%)	**2** (7%)	**1** (4%)		27
Inflammation	39	6	2	1	48
Hyperplasia	24		1	1	26
undetermined/other	23			2	25
total	**179** (84.4%)	**18** (8.5%)	**5** (2.4%)	**7** (3.3%)	**209**

for deferral was specified. In 7 more cases, the specialist did not give a qualified diagnosis but rather pointed out some possible differential diagnoses.

Table 11.2 summarises all 14 consultations which either showed a marked discrepancy between the TP diagnosis and the review or for which the value of the TP diagnosis was not sufficient for treatment. The table summarises the preliminary diagnosis by the physician at SHCH, the TP diagnosis and the review diagnosis. The comments were provided by the reviewing pathologist. The reviewer also specified a value of the TP diagnosis (sufficient for treatment, insufficient for treatment, wrong or other) and tried to specify a reason for diagnostic discrepancy.

In two cases the TP diagnosis was wrong or misleading and in 6 cases the TP diagnosis was insufficient for treatment. The discrepancy in case 1 is graded as "marked" because the TP diagnosis included a carcinoma as differential diagnosis which was not confirmed by the review. Case 2 is a typical example of communication failure – in the original submission only one image was included, probably due to a transmission error of email, and the pathologist gave a preliminary diagnosis on this one image on which an adenocarcinoma was not visible and asked for additional images. When the full set of images was submitted a few days later, the pathologist did not review the case, although on these images the adenocarcinoma was clearly visible. Cases 4 and 8 are other examples of failure of communication between case submitter and pathologist. Case 14 included three x-ray images and the pathologist wanted to consult a radiologist before giving a final diagnosis, but failed to do so.

In addition to case 2, there were four other cases with a false negative diagnosis. In case 3, the squamous cell carcinoma was not visible on the originally submitted images, although in the review islands of squamous cell carcinoma were found. However, the pathologist recommended a re-biopsy which was subsequently done and on which the carcinoma was diagnosed by TP and thus the initial false negative diagnosis had no impact on the patient. In case 9 the TP diagnosis was "non-specific inflammation" while the review revealed a erythrema nodosum. In cases 12 and 13 the TP diagnosis was deferred by the pathologist on duty and the cases were not forwarded to the expert group although a diagnosis would have been possible based on the submitted images.

Table 11.2.: This table gives an overview of the problematic cases in which either the value of the telepathology diagnosis was judged as misleading or insufficient for treatment or where the TP diagnosis was markedly discrepant from the review diagnosis.

	SHCH	TP	Review	Comments	Value	Reason	Discrepancy
1	keratosis	DD: pseudoepitheliomatous hyperplasia vs. squamous cell carcinoma	skin ulcer with suppurative inflammation and pseudoepitheliomatous hyperplasia	complete excision recommended 1 slide HE	sufficient for treatment (correct)	Incorrect diagnosis either TP or Cambodia	marked
2	liver tumor DD. cholangiocarcinoma HCC colonic carcinoma	large bile duct with inflammation and periductal fibrosis	stroma-ride adenocarcinoma of gall bladder or extrahepatic bile ducts	1) first image was not representative for diagnosis (failure of submitter) 2) there was no response to a second image (failure of pathologists) series appropriate for diangnosis Beispiel 2/5.1 2 slides HE	misleading (wrong)	Inadequate selection of images for iPath	marked
3	chronic non-specific inflammation	no final diagnosis, granulation tissue, re-biopsy recommended	granulation tissue with islands of a squamous cell carcinoma	TP-diagnosis of squamous cell carcinoma in the recommended subsequent biopsy (s. nr. 03-0232) 2 slides (HE, PAS)	misleading (wrong)	sampling error	marked
4	giant cell tumor	DD: (without x-ray) aneurysmal bone cyst	giant cell tumor of bone	TP diagnosis is preliminary (without x-ray) final diagnosis (after addition of x-rays) was not made. 1 slide HE	insufficient for treatment (suspicious)	failure of communication	others
5	DD: Inflammatory bowel disease, M. Crohn, colonic tumor	active chronic colitis with ulcerations, suspicious of Crohn's disease	colonic ulcer; ischemie or drug-induced (NSAR?)	2 slides (HE/Pas)	insufficient for treatment (suspicious)	Incorrect diagnosis either TP or Cambodia	moderate
6	reactive inflammatory lymphadenopathy, clinically suspect T10	suspicious of NHL, no final diagnosis	non-specific chronic lymphadenitis with eosinophilia	4 slides (2xHE, 2xGiemsa)	insufficient for treatment (suspicious)	Inadequate selection of images for iPath	marked
7	suspect breast cancer	no diagnosis, insufficient preparation and preservation	invasive ductal carcinoma	1 slide HE	insufficient for treatment (suspicious)	Inadequate selection of images for iPath	marked
8	multiple lymphomas, suspect NHL, bone marrow involvement?	no final diagnosis due to poor image quality	no hints at bone marrow infiltration by HNL	E adequate selection but poor image quality (first series) I no response of TP-pathologists to a subsequent series of images (with better quality)	insufficient for treatment (suspicious)	other	marked
9		chronic non-specific inflammation	cutane vasculitis, e.g. erythema nodosum	1 slide HE	insufficient for treatment (suspicious)	Incorrect diagnosis either TP or Cambodia	moderate
10		chronic active gastritis	medium-grade chronic Hp-gastritis of antral mucosa with moderate activity	helicobater pylori can be proved in the Giemsa-stain, not considered in the selection of images 4 slides (2xHE, 2xGiemsa)	other	other	minor
11		chronic gastritis with granulocytic activity	medium-grade chronic Hp-gastritis of antral mucosa with moderate activity	helicobater pylori can be demonstrated in Giemsastain, not considered in selection of images 4 slides (2xHE, 2xGiemsa)	other	other	minor
12	lymph node tuberculosis	no final diagnosis	hyalinized and fibrosized granulomatous lymphadenitis. TB (Z.-N. positive)	no final diagnosis (after demonstration of AFB in Ziehl-N. staining by the submitter) 2 slides HE	other	other	others (marked)
13	lymphoma infiltration, AML or ALL	image quality not sufficient for hematological diagnosis	suspicious of AML	Image quality not sufficient for TP-diagnosis, for diagnosis see 03-0141 AML 2 slides (HE, Giemsa)	other	other	others (marked)
14	giant cell tumor	giant cell tumor of bone with a secondary aneurysmic bone cyst	aneurysmic bone cyst	specimen is representative for an aneurysmic bone cyst, the diagnosis of a giant cell tumor was made on a former biopsy 5 slides HE	other	other	marked

11.3.2. Reasons for Diagnostic Discrepancies

In table 11.2 the reasons identified by the reviewing pathologist are listed. The most prominent reasons are inadequate selection of images and failure of communication. "Incorrect diagnosis either TP or Cambodia" indicates that no other explanation than a failure on the side of the pathologist could be identified.

To analyse possible reasons for diagnostic discrepancies, the following four characteristics were evaluated for all 212 cases: *tissue sampling*, *quality of histology*, *image selection* and *communication*. Table 11.3 summarises the results of this analysis. Tissue sampling and histological quality of the slides do not seem to influence the diagnostic discrepancies. The multi-field χ^2 test did not show any significant correlation. Communication problems and especially problems with image selection show a different picture and occur more often in cases with marked discrepancies between TP and review – χ^2 test shows significant correlation (p<0.001).

11.4. Discussion

The results of the study show a very high accuracy of the TP diagnosis provided. The TP diagnoses differed markedly from the review diagnoses based on the original glass slide in only 3.3% of the 212 cases. 84.4% of all cases showed complete concordance between TP and review. These figures are comparable to figures from other evaluations of static image telepathology [38, 163].

Under the given circumstances, telepathology may very well serve as a diagnostic tool for hospitals in developing countries that do not have a fully qualified resident pathologist.

The major problems observed are image selection and communication. Image selection is most directly related to telemedicine as under normal condition, there is no need to pre-select images – the pathologist can view the whole slide. Communication problems exist as well in conventional pathology where the clinicians sometimes fail to indicate important clinical facts to the pathologist. However, it is important to note that both these factors can easily be improved by training and experience.

An important lesson learned for the future development of the iPath and for any telemedicine software is to include better tools for organisation of communication. While issue concerning image quality and reliabilty of transmission are hardly observerd, the software should activly support the flow of communication. A possibility would be to offer specialists an option that they want to be automatically reminded to get back to a certain case after some defined time. If they decide that a case needs further input or clinical data or if they want to consult e.g. with a radiologists, then the system could notify them after a few hours or days reminding them that they wanted to further comment on this particular case. An extension of the workflow module on iPath that will include such reminder funtionality is certainly high on the agenda for the future.

Besides the benefit of access to timely diagnosis from an expert pathologist, the presented set-up has yet another very important aspect. The continuous interaction with expert pathologists offers

Table 11.3.: Reasons for discrepant diagnosis.

Tissue Sampling	total	identical	minor	moderate	marked	other
representative	*153*	132	14	3	2	2
representative for diagnosis, not sufficient samples	*40*	31	4	2	2	1
not representative	*5*	4			1	
no statement/other	*14*	12			2	

$\chi^2 = 15.98$ (not significant)

Quality of Histology	total	identical	minor	moderate	marked	other
very good	*3*	3				
good	*153*	135	11	3	2	2
moderate	*42*	29	6	2	4	1
insufficient	*10*	9	1			
no statement/other	*4*	3			1	

$\chi^2 = 19.15$ (not significant)

Image Selection	total	identical	minor	moderate	marked	other
all relevant changes documented	*154*	143	7	2		2
appropriate for diagnosis, some findings missing	*15*	7	7	1		
moderate in quality or quantity but sufficient for diagnosis	*14*	12	2			
insufficient	*10*	5			4	1
other	*19*	12	2	2	3	

$\chi^2 = 106.79$ (p<0.001)

Communication	total	identical	minor	moderate	marked	other
efficient	*175*	151	13	5	4	2
deficiency on case submitter's side	*16*	13 (2)	3			
deficiency on both sides	*3*				2 (2)	1 (1)
other	*11*	8	2		1	

(values in brackets indicate major communication problems)
$\chi^2 = 65.94$ (p<0.001)

an excellent teaching opportunity for the physicians doing the preparation of the tissue at SHCH. This is immensly important in a country like Cambodia, where after the civil war almost no senior specialist was left in the country. Young doctors hardly have the opportunity of asking a "next-door senior expert" for advice on a difficult case. In contrast to workshops and seminars and also to distance education programmes that alleviate some of these issues, telemedicine offers the possibility to discuss the very problems that occur at SHCH. The iPath telemedicine platform also allows the consultants to add additional files such as reference images or scientific articles to further illustrate the problem discussed. This could also be considered as problem-based learning in its purest form.

This form of collaboration of different experts on the same problems also illustrates one of the advantages of using a proper telemedicine platform like iPath over consultations via plain email. Cases sent via email are sent on a personal basis and are only available to the expert to whom they were submitted. If several experts are approached, their opinions are not collected in an organised form and only exist in the inbox of the case submitter.

On a well-organised telemedicine platform cases can be assigned to the specialist according to sub-specialty. Since the telepathology network of the University of Basel is hosting many sub-specialty discussion groups such as for example a bone tumour working group or hematopathology and dermatology forums, it is relatively easy to refer interesting cases to the a proper specialist. Furthermore the different consultants can also communicate with each other and finally, since all comments are automatically archived into a database, it is at any time possible to retrieve older consultations. Over time this builds into a tremenduous archive of interesting material that could be readily used e.g. as base material for problem-based learning.

11.4.1. Way Forward

In other regions in South East Asia similar projects have been started based on the experience from this project. Consultations are regularely requested from Laos and Bangladesh; and a few pilot cases have been submitted from Myanmar and Vietnam. In addition, some provincial hospitals in Thailand signalled their interested in this form of telepathology in order to consult with the specialists in Bangkok.

A vision for the future is an independent South East Asian telepathology network with some distinct components: 1) diagnostic services for remote hospitals with an expert group or a virtual institute in form of a collaboration between local specialists and volunteering international experts, 2) a component for continuous medical education for the regional specialists which could consist of a forum for interaction between the regional pathologists with each other and with international experts as well as some kind of organised teaching activity with scheduled (tele-)presentations and lectures delivered by international experts.

In order to evaulate to what extent telemedicine may also serve for capacity building and skills development for the involved Cambodian physicians, a study to evaluate the impact of this form of telepathology on the diagnostic capabilities of the physicians at SHCH is in preparation.

Throughout the project the physicians at SHCH have always noted their working diagnosis for every case submitted to TP. The planned study will compare the results from the beginning of the project as presented here with the results from later years and observe if there is an improvement in working diagnosis by the submitting physician.

Finally, besides telepathology, the very same technology is also applicable for any other medical specialty and could equally serve other telemedicine projects in the region.

12. Teledermatology in the Transkei

Brauchli K (1, 5), O'Mahony D (2), Madikane P (3), Morris C (4), L Lagrange (5), Oberholzer M (6), Banach L(1)

1) Telemedicine Unit, Walter Sisulu University, Mthatha, South Africa
2) General Practitioner, Port St. Johns, South Africa
3) Primary Helath Care Centre, Tsilitwa, South Africa
4) Council for Industrial and Scientific Research, Pretoria, South Africa
5) Ceclilia Makiwane Hospital, Mdantsane (East London), South Africa
6) Department of Pathology, University of Basel, Switzerland

> Working Paper. Previewed for publication in the Journal of Telemedicine and Telecare or the Southafrican Medical Journal.

12.1. Introduction

The Transkei region of South Africa is an area of approximately 40'000 km^2 (3.6% of South Africa's surface) located on the East coast between East London and Port Edward. It is home of approximately 4 million people (almost 10% of South Africa's population). The Transkei used to be an "independent homeland" during the apartheid era and is now part of the Eastern Cape Province. It is one of the poorest regions in South African with high levels of socio-economic deprivation and a health status comparable to poorer regions in sub-Saharan Africa[99].

The region is characterised by poor road infrastructure, poor telephone communication and limited health facilities. Primary health care is provided mainly by nurses at state funded clinics, supported by general practitioners in the public and private sectors. Specialty services such as dermatology are only available at the Nelson Mandela Accademic Hospital which is associated with the medical faculty of the Walter Sisulu University (WSU) in Mthatha.

Due to high poverty levels, long distance travel is not affordable for most of the population. Hence, nurses in rural clinics and family practitioners have to diagnose and treat practically all dermatology problems. About 15% of all consultations in primary care have a dermatology component[116]. Access to specialist dermatologist advice is important as family practitioners have insufficient training in dermatology and as compared to specialists, their diagnosis and management are significantly poorer.

In this study we analyse the possibilities to improve access for patients to dermatological care and to improve family practitioner clinical skills using telemedicine. The study included two pilot sites in rural Transkei. The first site is a general practice in Port St Johns, a small provincial town on the east coast of South Africa. In the rural Port St Johns district, the population numbers about 75 000, the majority of which lives below the poverty line. In the last decade, the number of doctors in the district has varied between two and six.

The referral hospital at Mthatha is 100km distant but since 1998 has had no dermatologist (except for a 2 year secondment of a Cuban specialist). The nearest dermatologist was in East London, 350km away. Teledermatology in Port St. Johns was started in 1999 with email based store-and-forward teledermatology[116], and since 2002 it is using iPath. In the first year the server in Basel was used but since 2003 the Telemedicine Unit of the Walter Sisulu University (WSU, former UNITRA) is running a regional telemedicine network[1] based on the iPath software[22] which is now being used by the teledermatology project in Port St. Johns.

The second site is the community clinic in Tsilitwa which is located about 120km from Mthatha. The clinic is serving a population of approximately 10'000 people. Tsilitwa has been a pilot site for a wireless telemedicine project initiated by the Council for Scientific Research (CSIR) which enabled the nurse at Tsilitwa clinic to consult with the GP at the Sulenkama Hospital over a distance of +- 30 km[105].

12.2. Material and Methods

12.2.1. Port St. Johns

This study presents an extension of the tele-dermatology study started in Port St. Johns in 1999 which analysed 52 teleconsultation taking part from April 1999 to December 2000[116]. Since then the number of consultations has grown to 110 consultations and includes consultation until December 2004.

Patients who presented at the family practice in Port St. Johns with dermatological problems were included in this study if the general practitioner (GP) could not make a diagnosis or wanted to have his preliminary diagnosis confirmed by a dermatologist. For the telemedical consultations images were captured with a digital camera (first an Olympus C-1400XL and later an Fuji 2 mega-pixel). If necessary, images were cropped and resized to approximately 1000x700 pixels

---

[1] http://telemed.utr.ac.za

using Adobe Photoshop or GIMP[2] and saved in JPEG format. In the beginning images were submitted by email with attached pictures. However, text and pictures and diagnostic replies easily got separated and misfiled. Thus patient information and images were compiled into an html page which worked well but was a very time consuming process. Since 2002 the iPath telemedicine platform was used for teledermatology consultations. With iPath, images and clinical description can beuploaded via an easy to use web site. Alternatively, cases can also be sent via email which are automatically imported into the same web based database (c.f. fig.12.3). The dermatologists are notified about a new case by email automatically and can review the case on the website. Any diagnosis or comment that they enter is immediatly sent to the case submitter via email. All data is achieved on the website so that it is available anytime for reference.

Since no web site can be entirely secured, no patient names are used for identification. The submitting practitioner must assign a numeric code to each consultation which only he can directly link to the real patient.

Initially, teledermatological consultations were submitted to Medical University of South Africa (MEDUNSA) in Pretoria, to the Armed Forces Institute of Pathology (AFIP) in Washington, USA and Lemuel Shattuck Hospital (LSH) in Boston MA, USA. In the second phase using the iPath platform of the University of Transkei (UNITRA), cases were reviewed by dermatologists at Mthatha General Hospital, Cecilia Makiwane Hospital in East London and by the dermatology department of University of Cape Town. With the web based solution of iPath, the task of reviewing cases could be easily shared between different dermatologists without the need of the case submitter to send multiple consultations.

12.2.2. Tsilitwa

Tsilitwa is a community located in a deep rural area of the Eastern Cape about 120km from Umtata and is a typical example of many of South Africa's rural areas that exist below subsistence levels and remain impoverished because they have no access to basic infrastructure essential for economic growth and development. As a consequence, the youth are leaving their rural homes in pursuit of employment and opportunity in the cities. Basic infrastructure such as electrical power, tarred roads and communications, essential pillars for economic growth, are often not available.

Tsilitwa is a pilot site for a project of the South African Council for Industrial and Scientific Research (CSIR) which aims to develop and implement an innovative communications infrastructure that is independent of the state power and telecommunication utility companies and to develop capacity within the Community to sustain and utilize this network. The project has created a wireless network which connects various institutions within the district. One of the pilot applications of this communication networks is a telemedicine link which facilitates the teleconsultation between a patient at the Tsilitwa clinic with the doctor at Sulenkama hospital using an Voice over IP (VoIP) telephone and a video camera.

---

[2]http://www.gimp.org/

Figure 12.1.: The community clinic in Tsilitwa. Images captured with the video camera can be transmitted live over the wireless IP network (WiFi) to the Sulenkama Hospital, approximately 30km away.

While this form of consultations are suitable for many types of medical quesions it cannot help with questions that require a medical specialist such as e.g. a dermatologist. Live consultation with dermatologists are not an option for two main reasons: 1) it is not possible to time patient and dermatologists at the same time and 2) the network does not have bandwidth to to video link outside of the WLAN. However, the clinic in Tsilitwa had all equipment necessary for store-and-forward tele-dermatology: a battery powered HP PhotoSmart camera with a fixed lense and a resolution of 2 MegaPixels resolution, a PC and an email connection to the outside network. Unfortunately, the email connectivity was discontinued at the beginning of the teledermatology study and thus only very few cases could be included in this study – these consultations were transported on memory stick form Tsilitwa to Mthatha and there uploaded to the UNITRA telemedicine platform.

12.2.3. The UNITRA Telemedicine Network

In the initial phase of the teledermatology project in Port St. Johns the consultations were submitted by plain email directly to the consultant. This works fine in a small group of people but it seemed not very feasible for extending the possibilities for teledermatology consultations to other primary care clinics within the Eastern Cape province. In 2002 the project tested the iPath server at the University of Basel and found this solution highly suitable as platform for teledermatology in low resource settings.

In order to prepare for a deployment of teledermatology at a provincial level it was decided to create a regional telemedicine network based on the iPath software. The regional network should ensure that all medical data will stay in South Africa and that the technical skills for maintaining and supporting the application will be available within the region.

A first server was established at the UNITRA telemedicine unit in June 2003. It was based first on an old Dell Optiplex PC hardware (Pentium II, 64MB memory) and was later migrated to a HP proliant server (P-IV, 256MB, 80GB HD). The server is running on Debian Linux and provides the following services:

- Content Management System (CMS) for the telemedicine unit website announcing news, conferences etc.
- E-mail server providing local email accounts for all participants.
- Telemedicine service based on the iPath software.

The email server was installed because it was found that most users were using a US based webmail service. For submitting telemedical consultations with attached images this was often very slow and it is a waste of limited international bandwidth[53].

12.3. Results

Since 1999, 110 patients from Port St. Johns have been diagnosed using teledermatology. 76 patients where female and 34 male with an average age of 32 years, ranging from 2 weeks to 73 years. 27 consultation were submitted to MEDUNSA, 10 to AFIP, 35 to LSH and 38 to UNITRA. Teledermatological assessment was possible in 105 cases. No definitive diagnosis was given in five cases. The diagnoses are summarised in table 12.1. For the whole The turnaround time for tele-diagnosis ranged from 1 day to 246 days (median = 11 days). The turnaround times improved considerable during the project and for the last 38 cases submitted via the UNITRA telemedicine platform all consultations were answered within one month (median = 8 days).

Figure 12.2 gives an overview of the results of the consultations. In 105 cases a diagnostic reply was possible. In 46 cases the dermatologists confirmed the provisional diagnosis of the GP. In 57 cases the telemedical diagnosis helped the GP to improve the treatment of the patient. The GP perceived these consultation as useful assistance in 104 cases. 72 of the patients returned to a follow up consultation to receive the results of the tele-consultations. For 36 patients a biopsy was taken and sent for histological diagnosis. In one case the histological diagnosis was not in accordance with the final diagnosis of the dermatologist. The preliminary diagnosis by the GP

Table 12.1.: Diagnoses made by teledermatology.

Diagnosis	N	Diagnosis	N
Tinea	10	Fibroepithelioma	1
Lichen planus	7	Fixed drug eruption	1
Erythema multiforme	4	Granuloma inguinale	1
Seborrhoeic dermatitis	4	Granulomatous lesion	1
Contact dermatitis	3	Haemangioma	1
Nummular eczema	3	Hairy leukoplakia	1
Pityriasis rosea	3	Herpes simplex	1
Psoriasis	3	Hyperkeratotic eczema	1
SLE	3	Infection	1
AIDS associated papulo-pruritic eruption	2	Keratoacanthoma	1
Atopic dermatitis	2	Keratosis pilaris	1
DLE	2	Lichenoid dermatitis	1
Granuloma annulare	2	Melanoma	1
Morphea	2	Metastatic cancer	1
Non-specific	2	Molloscum contagiosum	1
Ochronosis	2	None	1
Photosensitive dermatitis	2	Normal pigmentation	1
Scleroderma	2	Papular eczema	1
Seborrhoeic eczema	2	Pemphigoid	1
Acne	1	Pemphigus	1
Actinic chelitis	1	Peri-oral dermatitis	1
Actinomycosis	1	Pitted keratolysis	1
Alopecia areata	1	Pityriasiform eczema	1
Bacterial infection	1	Pityriasis rubra pilaris	1
Bullous lichen planus	1	POD	1
Candida infection	1	Pomade acne	1
Dermatitis herpetiformis	1	Pompholyx	1
Drug reaction	1	Possible syphilis	1
Dyshidrotic eczema	1	Pustular psoriasis	1
Eczema	1	Pyogenic granuloma	1
Epidermolysis bullosa	1	Subcorneal pustular dermatosis	1
Erythema annulare centrifugum	1	Warts	1

$n=105$

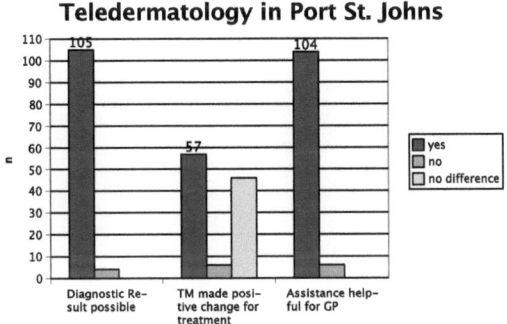

Figure 12.2.: Results of teledermatology consultations in Port St. Johns.

in this particular case was a Seborrhoeic eczema which was changed by the dermatologists to Pityriasis rosea. The histological findings of the biopsy were that of lichen planus.

From the clinic in Tsilitwa only 4 consultation could be performed. Three of the were diagnosed within 4 days. In one case a diagnosis was not possible due to missing clinical information. These cases give an impressive example of the potential of telemedicine for rural clinics. The history of the case presented in figure 12.3 is that the patient, a lady born in Tsilitwa and now working as an attendant at a petrol station in Johannesburg, started to develop itchy patches on the skin, especially in the face, which left ugly dark scares. She had consulted two doctors in Johannesburg and was given medication but without relief. Finally she lost her job due to her disease and came back her home village. When she started to develop symptoms of depression the nurse at the clinic in Tsilitwa decided to consult a dermatologist. She took the photographs which were then submitted to the telemedicine server at WSU. Two days later the dermatologist in East London reviewed the case and diagnosed it as Lupus Erythematosus. She suggested treatment with steroids and most importantly application of sun screen (zinc cream) as the skin deformation is caused by sunlight. A few weeks after this consultation the patient had returned to work.

12.4. Discussion

In general, the technology was working very well. The capturing of images with a digital camera was no problem for the family practitioner in Port St. Johns nor for the nurse in Tsilitwa. All

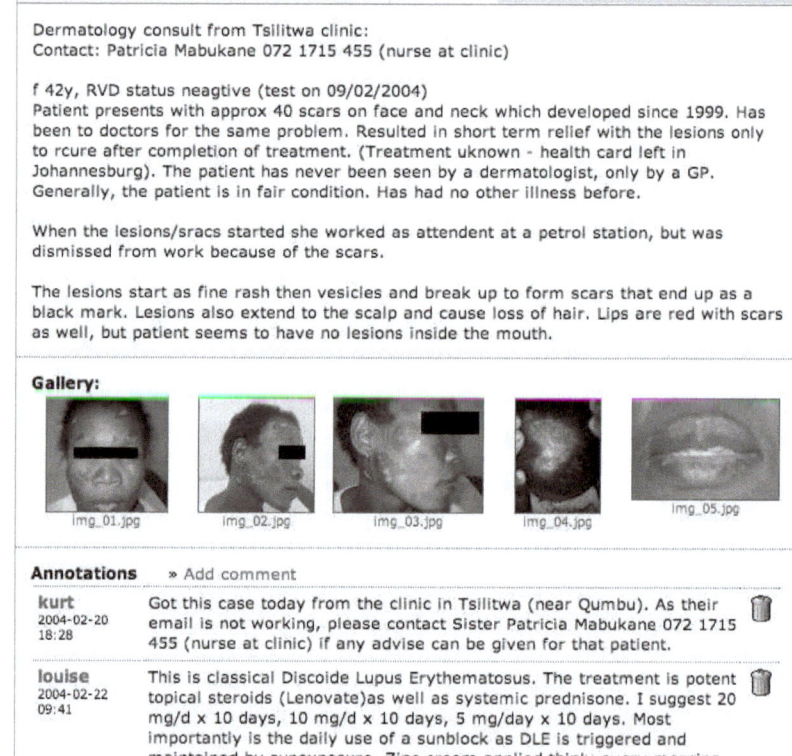

Figure 12.3.: Sample case submission of a dermatology consultation from the clinic in Tsilitwa. Description is given in the text.

images were accepted by the dermatologists as adequate for consultation. The transmission of images form Port St. Johns was no problem. Images were lost on one occasion were a consultation was submitted to the UNITRA web server after it had been moved from one hardware to another. From Tsilitwa electronic transmission was unfortunatly not possible after the GSM modem connection between their wireless LAN and the Internet was discontinued.

Though the project with Tsilitwa has not yet been very successful, it served very well as a role model for the potential of teledermatology in rural primary health clinics. If a minimal communication infrastructure consisting of a telephone connection and a personal computer is present a simple digital photo camera is all that is necessary to start teledermatology. The only weakness identified in the case study in Tsilitwa was that the nurse was not used in writing down the clinical history about the patient. In 3 out of 4 consultations the clinical history was judged as insufficient. The Department of Health of the Eastern Cape province is planning to equip ten rural clinics with a camera and to organise a workshop for the nurses where they will be taught a dermatologists how to take pictures of a patient and how to write a short clinical history relevant for a teledermatology consultation[3].

At the beginning of the project the turn around time for receiving dermatology consultations was often very long. In the initial year of the study 35% of the cases needed more than one month[116]. The reasons for this were manifold. The South African academic centres have a chronical shortness of medical specialists and especially in the initial phase the consultants were not yet accustomed to electronic communication. At some accademic centres access to internet and email is very restricted and some of the consultants reviewed the cases at home over the weekend.

However, the delays could be considerably reduced during the project and since the introduction of the iPath platform, it has become possible to share the diagnostic load between different consultants. On the web based platform the dermatologists can see if a case has already been reviewed by another dermatologist. In fact, the turn around time for the 23 cases that were submitted after the end of this study, the average turn around time has reduced again and is not less then 3 days.

The consultations played an important role in improving the treatment of the patients. In over 50% of the cases the diagnosis of the dermatologist changed the original treatment plan. The study was not directly looking into economical benefits of the consultations, it should however be noted that most of the patients that visit the family practise in Port St. Johns can hardly afford the 350km travel to the nearest dermatologist at Cecilia Makiwane Hospital in East London. Studies in the UK have demonstrated that teledermatology consultations are cost effective if the distance between the primary health centre and the specialist is more than 78km[170].

A major outcome however is not only the correct diagnosis and treatment for the patient but the fact that the GP strongly felt that teledermatology improved his skills in diagnosing and treating dermatology problems appropriately. The frequency of consultations has decreased towards the end of the project because one of his main objectives of learning was achieved. Even though

---

[3]this project was presented at the eHealth Indaba 2006 in East London.

in 42% of all consultations the diagnosis of the specialists was identical with the provisional diagnosis of the GP, he regarded it is a good learning experience to have his diagnoses confirmed. In the long term the number of cases will drop off – most patient visiting the family practise in Port St. Johns will receive a correct diagnosis and appropriate treatment directly from the family practitioners whose dermatological skill have considerably improved due to the regular tele-dermatology consultations.

Acknowledgement

This work was partially supported by a grant from the Fogarty International Centre of the National Institute of Health (I D43 TW01083-1) and by a personal grant of the Swiss National Science Foundation.

13. ICT for Distant Medical Collaboration in the Ukraine Swiss Perinatal Health Project

Blunier M (1*), Zahorulko T (2), Dobryanskyy D (3), Brauchli K (4)

1) Swiss Centre for International Health, Swiss Tropical Institute, Basel, Switzerland
2) Volyn Regional Children's Territorial Medical Centre, Lutsk, Ukraine
3) Danylo Halyts'kyy L'viv National Medical University, L'viv, Ukraine
4) Department of Pathology, University of Basel, Switzerland

*Correspondence address: marc.blunier@unibas.ch

This paper has been published in the Ukrainian Journal of Telemedicine, March 2006. [11]

Objectives: The Ukraine Swiss Perinatal Health Project aims to contribute to health system development by improving access to information and fostering professional networks for Ukrainian health professionals.
Methods: An internet based telemedicine platform was chosen enabling distant collaboration. Users were equipped with necessary infrastructure, basic PC, digital camera and internet access.
Results: 80 users from Ukraine and abroad registered to the network. 124 cases were presented in the network, 81 with a distinct clinical question, among them 68 received comments. Interaction took place between Ukrainian (case presentations) and international (providing comments) health professionals.
Discussion: It is shown, that ICT in health foster communication among professionals and contribute to continuous education over distance.

Keywords: Information & Communication Technology (ICT), Telemedicine, distant collaboration, knowledge sharing, health system development

13.1. Introduction

Access to up-to-date information is one pre-requisite to make informed decisions in all aspects of life. Information and communication technologies offer great potential to improve health services and systems.[174] Healthcare is about knowledge management, the right knowledge available at the right time in the right place directly influences the right outcomes. In transitional countries access to information is still compromised for various reasons. The direct personal exchange of knowledge and expertise among colleagues on a larger scale is hampered by long distances and slow public transport systems. Paper based media like medical literature and scientific papers are scarce due to limited financial resources and lacking distribution systems. Libraries provide a restricted choice of medical books not representing the full range of current and international recognised knowledge. Latest developments in medical science discussed in international journals, influencing clinical practice and leading to the modification of the clinical practice guidelines do not reach the attention of the medical community, especially not to those health professionals working outside the urban centres.

The United Nations World Summit on the Information Society (WSIS) 2003 in Geneva emphasised that access to Information and Communication Technology (ICT) is one of the main elements for the development of societies. For most of the industrialised countries the use of Information and Communication Technology (ICT) has emerged as a key to drive efficiency and effectiveness of their health systems.

The objective of the Ukraine Swiss Perinatal Health Project is to contribute to the health system development and to improve offer, quality and access to preventive and curative perinatal public health services in selected Ukrainian regions. The ICT component of the project aims to contribute to the overall objective by improving access to information for health professionals in Ukraine, fostering professional networks and establishing a Telemedicine platform through which information can be exchanged. It was expected that participants share their experience from daily clinical practice and provide expertise to answer medical questions of their colleagues, however there were no rules defined on how the network shall be used. Therefore participants applied the network very individually. This reflects also in the results section, where the activities are analysed from a general perspective and from the perspective of one district hospital. The involved health professionals are Obstetricians, Gynaecologists and Neonatologists from Ukraine and international colleagues. The technical infrastructure applied consists of personal computers, internet access, and a server application called iPath[16, 20]. iPath has been chosen because of its intuitive user interface, its moderate computer and communication infrastructure requirements and because it is developed in open source. Open source software is interesting for applications in low resource environments, where the costs for software and licences compromise the scarce health budget. Additionally, the open source model allows technology transfer and active participation by the Ukrainian partners also on the technical level. The technical infrastructure enables the participants to collaborate in a closed user group and to communicate, using a common platform and disposing of a structured format with functions to exchange data and information.

13.2. Materials and Methods

The project started in April 2003 with two partner hospitals recruited from the Ukraine Swiss Perinatal Health Project (USPHP). The Institute for Paediatrics, Obstetrics and Gynaecology in Kiev and the Regional Children Hospital of Ivano-Frankivsk appointed each a doctor to coordinate the activities within the hospital. The coordinators received training for the application of iPath and the digital camera. They have been taught how to enter medical cases into the database and add digital files, how to open existing medical cases and how to provide comments. Interested doctors from these hospitals can refer to the coordinators to get information and trained in order to participate in the network. The department of Obstetrics and Gynaecology of the University Hospital of Zurich is the international counterpart within the network.

To facilitate tele-consultations as well as sharing and discussion of clinical information, the project selected the iPath telemedicine platform for implementing the Ukraine-Swiss perinatal telemedicine network. iPath is a web-based, open source telemedicine platform developed at the University of Basel since 2001[16, 18, 20, 22]. The iPath platform combines communication with content management features and its main function is the "medical discussion group" in which a defined group of users can present and discuss clinical cases.

iPath provides the user with a structured format to present the cases (cf. fig.13.1). The first section contains the main information about the presented case e.g. patient information, anamneses and diagnosis in plain text. The sender explains the reason for presenting the case and formulates question he/she wants to discuss. To the gallery the users can add images and any other documents. The annotation section lists the comments provided by the users. Comments are entered directly from the web or can be sent by email. Comments via email are automatically integrated into the case.

Instead of operating its own server, the project started with utilizing the existing iPath server at University of Basel, operated by the Department of Pathology of the University of Basel[1]. This server is used worldwide by over 1000 users and is also hosting many telemedicine applications with developing countries[22]. Since April 2003 the network was growing continuously and new participants added to the network like the six partner hospitals from the Perinatal Health Project and individual doctors working in other hospitals. Up today 80 individual doctors (61 Ukrainian and 19 international) registered to iPath and are members of the network.

The USPHP is operating its own discussion group on iPath. User first register a user account on the iPath server and can then apply to become member of the Ukraine Swiss Perinatal Health Group. iPath is accessible from any personal computer with internet connection. The most important functions are also accessible through an email interface, which is particularly useful for participants with slow connectivity and for receiving automatic notifications about new cases and comments.

The technical pre-conditions in the partner hospitals were varying from no infrastructure available to fully equipped and operational computer work places. The Perinatal Health Project pro-

---
[1] http://telemed.ipath.ch/

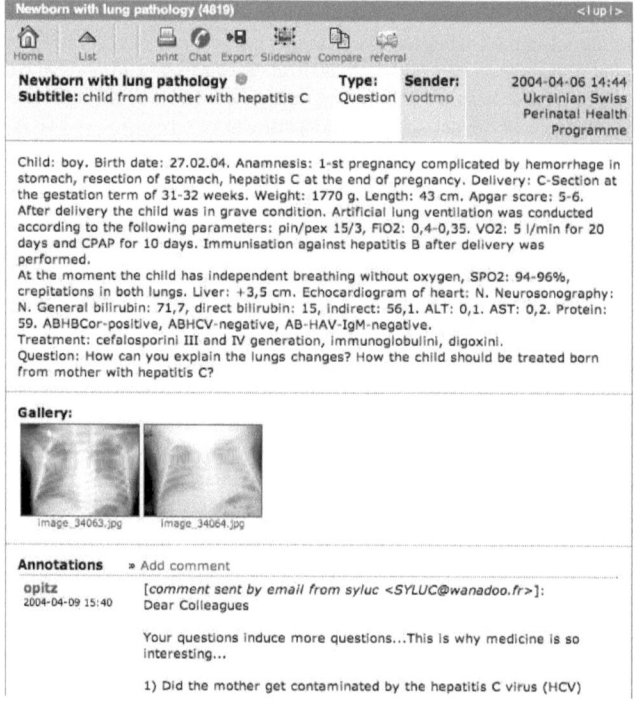

Figure 13.1.: This figure represents a typical case presentation on iPath. The header gives information about the type of case, submission date and the submitting user. This is followed by a textual description of the case which should include relevant clinical information and which should clearly formulate a question that should be discussed. Additionally the case can be illustrated with digital images or other documents. Cased details can only be modified by the owner of the case; however, other group members can add comments at the bottom of the case.

vided the needed infrastructure where necessary to the partner hospitals. Participants outside the partner hospitals use their professional or private technical infrastructure. The required equipment for a working place consists of a desktop computer (minimum Pentium III/1 GHz or better, 256MB, 60GB HDD), monitor (17" CRT with XGA resolution - 1024x768 pixels), flat bed scanner (1200dpi) and a digital camera. The cameras should have a resolution of 2 mega pixels or better, a 3x optical zoom lens with macro mode and the option to switch of the flash. For the project, Pentax Optio 33L cameras were chosen.

The computer work places are connected to the internet through digital modem connections of minimum 64kB/s, either ISDN or DSL. The analogue dial-up modem connections available in some partner hospitals were not sufficient in terms of speed, availability and reliability due to the bad physical quality of the telephone lines. Occasionally no separate physical line was available or it was connected through the hospital's telephone switch board disabling an internet connection. For those cases a separate physical line was installed between the hospital and the nearest point of presence of an internet service provider. Internet connectivity was established through various local internet service providers.

13.3. Results

The clinical cases entered into ipath between September 2003 and January 2006, were considered in the study. The study analysed the way Information and Communication Technologies (ICTs) did enable the development of the projected professional network and how the technology was adopted by the user.

The following user and usage aspects have been analysed:

- Number of user and frequency of participation
- Characteristics of network participation
- Number and characteristics of cases presented in the network
- Level and Quality of Interaction within the network

13.3.1. User statistics

From the total number of 80 registrations 61 (76%) originate from Ukrainian and 19 (24%) from international health professionals. Health professionals summarises Obstetricians, Gynaecologists, Neonatologist and Paediatricians. 14 (17.5%) persons just completed registration but did never access the system - they are not considered to be users. 21 (32%) of the registered users logged into the network more than 25 times, whereas 45 (68%) registered users logged into the network less than 25 times. 11 users logged in more than 100 times.

	Obstetrics	Gynaecology	Neonatology	Paediatrics	Total
Presentations	22 *(11)*	7 *(0)*	8 *(5)*	6 *(2)*	43 *(18)*
Consultations	43 *(38)*	5 *(4)*	27 *(23)*	6 *(3)*	81 *(68)*
Total	65	12	35	12	124

Table 13.1.: This table describes the distribution of cases. The numbers in brackets indicate the number of cases that received at least one comment. Out of the 124 case, 43 were case presentation and 81 were consultations. From the 43 case presentations only 18 received any comments (42%). From the 81 consultations 68 received at least one comment (84%).

26 (39%) registered users logged into the network at least once during the time period of the last three months (11/05 to 01/06). 40 (61%) of the registered users have not logged in for more than 3 months.

There are active and non-active (only reading) users. 22 (33%) registered users opened at least one case for discussion and 29 (44%) registered users did provide one or more comments to presented cases. 18 (27%) users are providing both, cases and comments to the network. 34 (52%) users did never post a case or write any comment, 10 of these reading only users logged in more than 10 times into the system.

13.3.2. Case statistics

During the observation period 124 cases were opened. All the cases have been opened by Ukrainian users; international users did not open cases. The cases can be distinguished between pure clinical case presentations (n=43) and consultations, i.e. case presentations with distinct clinical questions (n=81). As table 13.1 illustrates, cases represented the fields of Obstetrics (52%), Gynaecology (10%), Neonatology (28%) and Paediatrics (10%).

36 (29%) of the cases contained text only and 88 (71%) contained images e.g. morphologic images of a body parts were contained in 42 cases, radiological images in 23 and ultrasound images in 48 cases. For the cases that were presented with images the number of images ranged from 1 to 25 with an average of 4.7 images per case.

13.3.3. Comment statistics

38 (31%) of the total of 124 cases did not receive any comment. 86 of the cases received one or more comments. From the 81 cases presented as consultations 13 (16%) did not receive any comment. From the total of 165 comments provided in the network 50 (30%) have been provided within three calendar days. 21 Ukrainian users have provided together 86 comments with an average of 4.1 comments per user; 7 international user together stand for 79 comments, on average 11.3 comments per user. Requests for clarifications or final outcome of the presented cases were issued 32 times, but only 13 times a reply was given.

users with most cases/session

Rank	Sessions	Cases	Cases/Session	Origin
1.	8	2	0.25	Ukraine
2.	10	2	0.2	Ukraine
3.	5	1	0.2	Ukraine
4.	75	14	0.19	Ukraine
5.	6	1	0.17	Ukraine

users with most comments/session

Rank	Sessions	Comments	Comments/Session	Origin
1.	20	11	0.55	Int.
2.	10	5	0.5	Ukraine
3.	96	35	0.36	Int.
4.	17	5	0.29	Int.
5.	69	15	0.22	Int.

Table 13.2.: This table illustrates the characteristics of user activity by estimating the ratio of cases and comments per session. The table shows the top 5 user who posted most cases/session and who posted most comments/session. Interestingly, the activity is very asymmetric. While the most active users posting cases are all clinicians from regional hospitals in Ukraine, the user most actively posting comments are mainly international participants. Interestingly, none of the very active users (100+ logins) is in either top 5 list.

To characterise the different usage of the system we set the activity (number of cases and number of comments) in relation with the number of sessions (logins). A high ratio of cases per session indicates that a user is mainly interested in using the telemedicine platform for presenting or consulting own cases. A high ratio of comments per session illustrates that the user's main interest is answering clinical questions. Table 13.2 illustrates that these two groups are clearly different. The users with the highest case/session ratio are all clinicians from regional hospitals. On the other hand, among the 5 users with the highest comments/session ratio 4 are international participants. Further it is interesting to know that none of the most active users with more than 100 logins are among either of these two groups.

13.4. Clinical Application

The Volyn Regional Children's Territorial Medical Centre used the platform mainly for distant medical consultation of their patients. The description of the patient history, the documentation of the clinical problem in writing and with images and the formulation of the clinical questions has been prepared in iPath. The team using iPath includes physicians of different specialities e.g. Neonatologists, Paediatricians, Surgeons, Obstetricians and Gynaecologists. Neonatologists

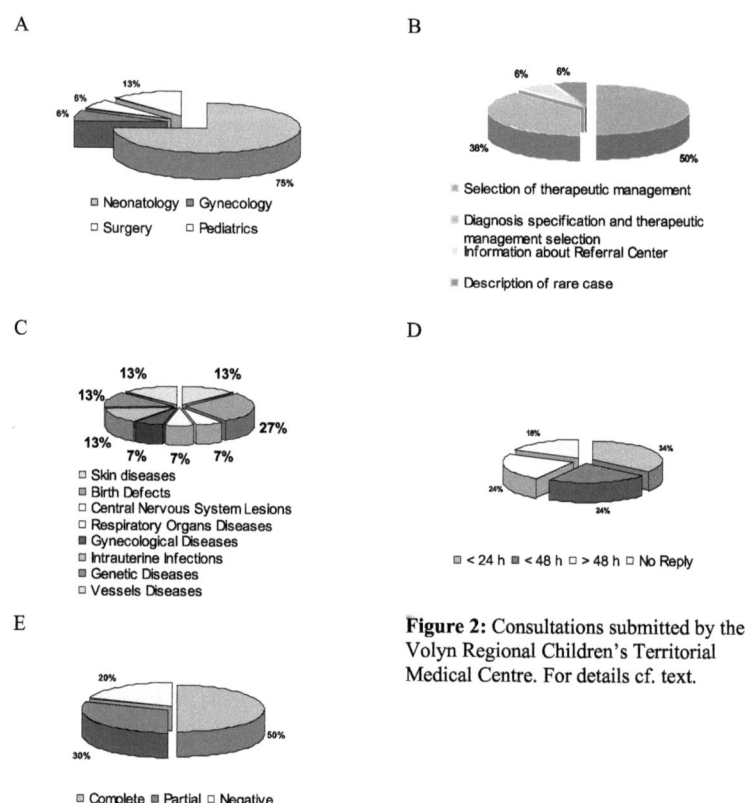

Figure 2: Consultations submitted by the Volyn Regional Children's Territorial Medical Centre. For details cf. text.

used iPath more frequently.

There have been 16 cases uploaded since September 2003 (fig.13.4A/B). The objective of the clinical integration of iPath was to obtain additional information from colleagues outside of the own working environment in order to make informed decisions. ECG data, X-Ray images, ultrasound and computer tomography images as well as laboratory examinations were uploaded and discussed using iPath.

Diagnosis specification and therapeutic management selection were needed for the patients with skin disease, genetic disease (facomatosis, achondroplasia). Patients with intrauterine infection, sepsis, hyperplasia of endometrium, bronchopulmonary displasia and central nervous system lesion needed the specification of therapeutic management. Additionally, a rare case of a child with birth defect of the brain (holoprosencephaly) was described. The information concerning an oppropriate choice of a Referral Centre was needed for a patient with mandibulofacial disostosis

(Franchescetti syndrome). The nosological structure of the uploaded cases was diverse as it can be seen in figure 13.4C.

The leading place is taken by congenital pathology as more hard and significant pathology for correction and treatment. After receiving comments a clinical meeting was organised, where all recommendations and suggestions were discussed among specialists and they decided the possibility to use them in this situation.

Most of the consultations were done instantly. This gave the possibility to make appropriate decisions e.g. to change antibacterial therapy, to correct artificial lung ventilation parameters, to provide differential diagnosis and additional examination of patients.

The time delay until the expected information is received is one important factor for the usefulness of this information. For 5 (34%) cases the information arrived within 24 hours, for 4 (24%) cases it arrived within 48 hours. For 3 (18%) cases no answer was provided within the network (fig.13.4D).Partial or complete satisfaction was given in the cases when recommendations could be applied in the clinical process. The counselling is considered to be negative in case of no reply because each request for information had a professional interest and a clinical relevance. The most efficient consultations resulted for the following cases: provision of artificial lung ventilation, antibacterial therapy, assessment of immunological results (fig.13.4E). Satisfaction of the consultation for patient can be hardly estimated as it depends on the final result of the treatment and this is not always objectively measurable.

13.5. Discussion

When the Ukrainian Swiss Perinatal Health Project introduced a telemedicine component in April 2003 the main goal of this initiative was to foster communication between Ukrainian specialists and the involved medical partners in Switzerland and Western Europe. The initial interest was high and 80 users had registered. However, only 45 users logged in regularly and only 26 users logged in during the last 3 months. On first sight this might seem as a very low rate of participation it actually corresponds very well with the general picture of such a telemedicine platform. Looking at the iPath telemedicine server of the University of Basel on which the project's application is hosted reveals a similar picture. Out of over 2100 registrations, only about 1000 users are actively using the server. For the Ukrainian partners in this project, especially those working in regional hospitals, the telemedicine tool provided an opportunity to overcome professional isolation and to share clinical experiences and consult difficult cases with specialists from different countries (Switzerland, Hungary, France, USA and Ukraine). While the initial goal of the telemedicine component was focused on the sharing and mutual presentation of interesting medical cases, it has become obvious that there is also a very strong demand for consultation. Out of the 124 cases submitted to the telemedicine platform, only 43 were case presentations. 81 cases were submitted as clinical consultations.

As table 13.2 illustrates, there was a distinguished difference in the participation pattern between the international and the Ukrainian participants in the network. Typically, the Ukrainian partners submitted consultations while the international partners are providing comments. Considering the fact that one single doctor submitted 50% of the comments from Ukrainian users, the input provided by Ukrainian experts was relatively low.

From the technical side, establishing communication links was not so much a problem. The initial technical problems with connectivity through modem connections could be relatively easily resolved by upgrading the telephone lines form the main hospital telephone system to a new telephone line installed by the internet service providers. However, technical connectivity alone is not sufficient for successful telemedicine consultations. 31% of the presented cases did not receive any answer and only 30% of the comments were submitted within 3 days from the presentation of the case. Out of the 32 cases in which further clinical details were requested, these were provided only for 13 cases. However, this seemingly low activity also reflects the fact that the application was from the beginning designed as a tool for information sharing rather than direct clinical consultations.

Additionally, some of the international consultants had been unwilling or unable to comment directly inside the platform. They accepted consultations by personal e-mail, but then only the one person sending the email would be able to read their comments and the large part of the group who was interested in this forum as a source of constant professional training would be deprived of this input. As a way out, some users started to post comments, which they received by personal email, into the platform to share these comments with their colleagues.

Some participants perceived the use of English for case discussions as a barrier for wider acceptance of the telemedicine platform among Ukrainian health professionals. As a consequence, some of the users from the Donetsk region founded an additional regional forum using Ukrainian language. In this group 25 cases have been presented and discussed in Ukrainian. Additionally, they also translated the user interface of iPath into Ukrainian language.

Based on the experiences summarised in this paper, the USPHP has come to the conclusion that the telemedicine component should be continued, but that a new way of organising the workflow is necessary. In order to find an appropriate form of organisation and to plan the expansion of the telemedicine component a "telemedicine working group" has been formed in Ukraine and this group is now taking the leading role in defining the future of the telemedicine component within the USPHP. The working group is currently planning the following steps:

1. To set up its own telemedicine server in Ukraine[2], using the open source iPath software. This server will then be available for all medical disciplines such as dermatology, pathology or radiology[22].

2. Organisation of a consultation service for clinical problems in the perinatal medicine. This service shall be organised in the form of a Virtual Institute such as the iPath project is already using e.g. in the telepathology projects[18].

3. To enhance the quality of the case descriptions and the communication the working group is preparing several consultation forms. These forms will allow the users to submit the

---

[2]The server in Ukraine is now installed and data will be transferred in June 2006.

description of their problem in a more structured and complete way. This should also have a positive impact on the number of consultants' responses. As the medical problems presented within this network are very diverse, a single case description form will not suffice.

The Ukraine partners are also interested in expanding the network to the district level and to organise interregional consultation groups. The integration of this consultancy services into the daily clinical process will require an organisational structure to ensure timely and relevant replies to clinical consultations. For such a structure, the involvement of additional local experts who are willing to provide their expertise is required. Additionally the group's activity has to be moderated from a health professional ensuring that presented cases are according to the standards set by the telemedicine working group and to ensure that cases are forwarded to the consultants.

Finally, a second goal of the telemedicine working group is to build up an organisation that will be able to continue independently the telemedicine server in Ukraine after termination of the USPHP project itself.

13.6. Conclusion

It has been shown that ICT is a valuable instrument to enable access to information and to foster the exchange of experience among national and international health professionals. It is contributing to the continuous professional education and may even increase the capacity of a health system. On the other hand it became obvious that technology alone, without the appropriate structures behind which organise the network, the benefits are not fully realised. The positive experiences made so far encourage the project to further develop the network.

Acknowledgements

The Ukraine Swiss Perinatal Health Project is financed by the Swiss Agency for Development and Cooperation, Berne, Switzerland. The authors are grateful for the opportunity to conduct this study in the frame of this project.

Part IV.

Consolidation of Results and Discussion

The iPath telemedice platform developed during this project has matured to a useful tool for the organisation of telemedical collaborations, not only in highly developed nations but particularly in areas with limited resources. Although the technical development of the iPath telemedicine platform and its continuous adaptation to new needs as well as support and instruction of its users presented the most time-consuming part of my PhD project, I would like to concentrate the discussion on the results of its applications. A first chapter will discuss the outcome and experiences gathered in the results presented above and will place the results in a health system context. So far the application of telemedicine in developing countries has been restricted mainly to individual small scale projects addressing the accurate need of an isolated institution of health providers. Chapter 15 will discuss if and how telemedicine can be used to strengthen health systems in resource-constrained areas in order to allow the population at large to benefit. Chapter 15.4 will look at issues involved with the large scale implementation of telemedicine at a health system level and finally, before coming to the conclusionary remarks, chapter 16 will give an outlook on on-going projects that have resulted from the activities presented in this thesis.

This whole thesis is focused to some extent on telepathology, the application of telemedicine in the field pathology. This is mainly due to the fact that most of the work was carried out at the Department of Pathology at the University of Basel and thus pathology is the main medical field into which I did get some detailed insight. However, many of the findings and principles are portable to other fields of medicine. Although telepathology in developing countries is sometimes criticised as superfluous because the impact of a pathology diagnosis is often restricted by the limited availability of therapeutic methods in poor countries, especially for treating malignant tumours, it should be taken into account that the remaining therapeutic options are often associated with drastic consequences[132] and that the decision is entirely up to the surgeon or the attending physician. Besides, pathology is important for the understanding the physiological mechanism of diseases and thus for the understanding the effects of available medications. Pathology provides fundamental inputs on clinical decision making, continuous education of physicians and finally it often provides the ultimate quality control for clinical diagnosis by other means.

14. Telemedicine in Resource-Constrained Areas

Based on the iPath telemedicine software, an openly accessible Internet server was installed at the University of Basel. Over time this server was used by an increasing number of groups for many different types of telemedical applications of which the most important have been described in chapter 7. We could observe different possible applications of iPath and gained valuable insight into certain mechanisms and problems in the application of telemedicine under different conditions.

14.1. Telepathology

In two projects – one in Solomon Islands and one in Cambodia (chapter 9-11) – we analysed the diagnostic accuracy and validity of this type of store-and-forward telemedicine in the field of clinical pathology. While both studies showed that with relatively simple means an acceptable level of diagnostic accuracy can be achieved there are also some remarkable differences. The telepathology diagnosis on submissions from Solomon Islands show a significantly higher percentage of discrepant diagnosis ($p<0.001$, $\chi^2 = 26.3$) than those from Cambodia. As table 14.1 illustrates, the difference was especially pronounced for diagnoses with moderated and marked discrepancies.

While the two laboratories cannot be compared in terms of technical possibilities – the Sihanouk Hospital (SHCH) in Cambodia has substantially better laboratory facilities than the very simple lab at the National Referral Hospital (NRH) on Solomon Islands – it is not very probable that the differences in diagnostic outcome can be explained be the different laboratory set-ups.

Table 14.1.: Diagnostic accuracy of telepathology

	identical	minor	moderate	marked
NRH	69.3%	12.3%	8%	10.3%
SHCH	84.4%	8.5%	2.4%	3.3%

Table 14.2.: The influence of different factors on diagnostic accuracy. The non-relevant presents the sum of identical, minor and moderate discrepancy. Clinically relevant are only the marked discrepancies. The figures give the percentage of specimen with insufficient tissue sampling, slide quality, etc. (c.f. text)

	Discrepancy	Tissue Sampling	Slide Quality	Image Selection	Communication
NRH	non-relevant	3%	3%	8%	11%
	relevant	19%	7%	*52%*	*41%*
	total	4.6%	3.5%	12.6%	14.6%
SHCH	non-relevant	2%	5%	2.5%	8%
	relevant	14%	0%	*57%*	*29%*
	total	2.5%	5%	4.5%	9%

Table 14.2 shows that the histological quality of the slides from the Solomons (3.5% with insufficient quality) was in general even better than those of the slides from the more sophisticated lab in Cambodia (5% insufficient). The most pronounced difference is encountered for the quality of image selection – in 12.6% of all submissions from the Solomons the selection was not representative while from Cambodia these were only 4.5%. This fact is very important as the selection of images has been found to have by far the strongest correlation with diagnostic accuracy in both studies (c.f. tab.10.4 and tab.11.3). Table 14.2 illustrates that non-relevant image selection was associated with over 50% of the clinically relevant (marked) discrepancies as compared to 8% resp. 2.5% for the total collective. This difference can only be explained by the referring physician's background and his ability to select relevant images for different pathological conditions.

14.1.1. The Role of the Referring Physician's Clinical Background

The referring physicians at NRH are all surgeons (general and orthopaedic) who do not have any special training in histopathology beyond a general histology course that is part of the curriculum for all medical doctors. It was clear from the beginning of the project that training and guidance on how to select relevant images had to be provided by the remote pathologists in some form. However, it was quite unclear to what extent this would be possible using remote collaboration only. Figure 10.4 illustrates that there was a remarkable increase in the quality of image selection over time. The percentage of submission with relevant image selection increased from 35% in the first semester to 69% in the last semester observed. Notably, one of the first changes of the iPath system after the first images from Honiara had arrived was to introduce a coordinate system that allowed the remote pathologists to refer to a certain area on low magnification images from which they would like to have more images at higher magnifications.

In contrast to the situation at the NRH, the submitting physician at the SHCH is a specialist for internal medicine, who had undergone a one-year training in pathology in Europe prior to starting with the telepathology activities. The percentage of case submissions with relevant image selection was 76%. An increase of the quality of image selection over time could not be observed. However, it must be noted that the SHCH study included a shorter time period than the NRH study. A further study including more recent case submissions from SHCH is under preparation and might reveal insight on the development of the quality of TP services over time (c.f. section 16.1). The different clinical backgrounds of the physicians was prbably also responsible for the sampling of tissue for histological examination (grossing). In the SHCH study, tissue sampling was graded inadequate in 2.5% of all submission. For the NRH study these were 4.6%.

Furthermore, the clinical background of the submitting physician determines the expectation about the content of the report by the pathologists. The surgeons from NRH soon complained when the pathologist started discussing an interesting case and asked for one concise answer rather than an academic dispute. The physician at SHCH generally did not object discussions as they provided interesting learning material. The other extreme could be observed in the "histopathology forum" (c.f. page 58), where cases are submitted by pathologist who often even positively acknowledged an extended discussion about possible differential diagnosis and further diagnostic possibilities.

14.1.2. Extending Existing Collaboration via Telemedicine

In both projects in Solomon Islands and Cambodia telemedicine has played a major role in improving collaboration within existing partnerships. In contrast to site visits and organised training courses, telemedicine is neither restricted by time nor by geographical location of the partners. This includes follow-ups after training courses, support of visiting specialists and students on exchange programs as well as clinical tele-consultations and distance education.

The virtual community model as implemented by the iPath platform has been found particularly useful in this respect. In contrast to email which is very useful for collaboration on a personal basis, the virtual communities on iPath facilitates collaboration in a larger group and multi-group collaboration. In these examples, the hospital in Cambodia had a collaboration with a pathologist from Aurich (Germany), Solomon Islands was collaborating with Swiss doctors and a pathologist in Bangladesh was in touch with colleagues in the UK and South Africa. Thanks to telemedicine it has become possible to organise a virtual "expert team" with participation from all these partners, which is now providing high quality expert advise to a number of projects. The organisation of such a "virtual institute" has also been very valuable for the delivery of timely diagnosis in more complicated cases, but it would not have been feasible for one single project only. A certain minimal amount of activity is essential before the benefits of a virtual institute outweighs the necessary organisation effort.

14.1.3. Educational Impact

The educational impact of continuous access to consultations and expert advise should not be underestimated. While an individual consultation may privide direct benefits only for a single patient, there is always some form of knowledge transfer associated with a consultation. Such knowledge transfer may potentially translate into a skills development for the involved physicians. The educational impact of telemedicine for health care practitioners was explicitly addressed in the study on a teledermatology project in South Africa (chapter 12) indicating that telemedicine can have a positive impact on skills development of the involved partners although the validity and extent of this possibility must certainly be subject to further studies (c.f. section 16.1).

14.2. Regional Networks

A second aspect extensively studied was the possibilities for and the importance of regional telemedicine networks. The benefits of telemedicine in the context of resource-constrained areas is often viewed only in the form of international collaborations – basically "outsourcing" medical services to (volunteering) foreign specialists. However, telemedicine can also provide powerful tools for regional networks by facilitating and improving collaboration between different levels of health care in a certain region. While telemedicine as tool for international collaboration with developing countries has received some attention[18, 50, 116, 148, 172], there are almost no reports on specific opportunities and challenges of utilizing telemedicine for improving regional collaboration within a health system in a resource-constrained area. However, if telemedicine is to provide a substantial impact on strengthening health systems, fostering regional collaboration is probably the most important aspect[53].

Nonetheless, it is more likely that telemedicine will start in form of international collaboration. However, if strengthening of a health care system is envisaged, telemedicine applications should be designed in a way to promote not only North-South collaboration but also South-North and most importantly South-South partnerships. If a project fails to include the regional medical specialists into the telemedicine activities, they may easily perceive telemedicine as an unfair competition to their own services and thus try to prevent its implementation on a larger scale.

When delivering telemedical advice to developing countries, it should not only be correct but also relevant. The therapeutic intervention recommended from Switzerland, based on latest evidence, may not be applicable in Solomon Islands due to financial constraints or unavailable infrastructure. Telemedicine is often utilised for making referral decisions. For a successful outcome it is important that a patient is referred to the most appropriate health care facility but preferably as close to his home as possible. In this situation the involvement of the regional medical specialists is vital.

Within the scope of this thesis two regional networks could be studied. The telemedicine server at the University of Transkei was implemented to serve as a technically independent platform for various telemedical collaborations within the Eastern Cape region (c.f. chapter 12). It was well accepted by the existing teledermatology project. The two dermatologists, who are working for the public health services of the Eastern Cape, could be easily motivate to use it. In addition, dermatologists form the University of Cape town also started to help with the consultations. Turn around times for dermatology consultations were substantially reduced after the introduction of the regional network. However, the network failed to develop significant activity beyond the teledermatology project. Although the initial interest was significant, it has not been possible to motivate a local clinical champion for any other discipline who would initially promote the idea. In addition it seemed that there was a strong competition between different technical promoters of telemedicine in South Africa. Various projects had been carried out, but technology had stayed mostly underutilised[60, 61]. A collaboration between the different technical projects on a national level was not established and it seemed that all projects were trying to keep their few clinical applications tied to their technology rather than investigating into ways of deploying telemedicine at a health system level. In 2003, the Department of Health of the Eastern Cape province was planning to use the UNITRA telemedicine platform for an extensive teledermatology network including initially 10-20 primary care clinics within the province. Unfortunately these plans have not yet been put into practice and it had thus not been possible to study the feasibility of such an approach.

A second regional network observed during this project was the telemedicine initiative of the Ukrainian Swiss Perinatal Health Project (c.f. chapter 13). Initially the project tried to integrate telemedicine into their program as a tool for connecting perinatal specialists in Ukraine with partnering specialists in Western Europe. In a first evaluation workshop it was found that the Ukrainian partners welcomed the input from the specialists from Western Europe, but they wished more participation form the Ukrainian medical specialists. To improve acceptance of the system, the user interface was translated into Ukrainian language and case discussions in Ukrainian were promoted. In 2006, a separate iPath server was established in collaboration with the Mohila Academy School of Public Health in Kiev. The iPath software was installed on their existing web server. After successful testing of this server, all data from the iPath-server in Basel (cases, comments and user accounts) were transferred to the new server in Kiev[1]. It is envisaged that the Ukrainian partners in the project utilise the telemedicine tool to improve national/regional collaboration and communication in the field of perinatal health and possibly other medical disciplines.

Technical problems for the installation of independent regional networks based on the iPath telemedicine platform did certainly existed, but they proved to be the lesser barrier to successful implementation than the organisation and participation of regional users. The web- and email-based platform running on a low cost Linux server has been technically very stable and even the frequent power failures and network outages at UNITRA did not substantially compromise the system. Within an existing organisational setting (e.g. University) it is relatively easy to set up and run such a platform. However, successful implementation of such a network largely depends on the local clinical champions and on a fruitful collaboration between clinical and technical

---

[1] http://ipath.ukma.kiev.ua/

partners. There must be at least one clinical partner who is willing to develop a fundamental understanding of the necessary organisational framework in order to be able to translate the technical possibilities into real clinical application.

Regional networks are most useful to implement telemedicine at the primary care level where language, cultural background and familiarity of the experts with the local situation are of particular importance. For the organisation of medical specialist networks, the limitation to a regional audience can also have a negative impact[25].

15. Telemedicine and Health Systems

Telemedicine has often been suggested as a possible solution to improve health care delivery in developing countries[52, 53, 168, 171]. It has been well-documented that telemedicine is a viable option for improving care in a wide range of medical specialties. The analyses in Part III demonstrate that telemedicine can deliver timely and highly accurate remote diagnoses and decision support in areas with limited resource where no local specialists are available. However, telemedicine has mainly been used in individual care and for isolated consultations between distant specialists. Available reports on the application of telemedicine in developing countries are mainly descriptions and analysis of small scale projects that improve care delivery for few individuals – usually with the help of volunteering foreign specialists[18, 22, 67, 70, 116, 140, 172]. If the population at large is to benefit from telemedicine, it must be integrated into the health system in a way that the whole health system and thus the population at large can benefit from it. Although an implementation at a system level is far beyond the limited scope of a PhD thesis, I would like to discuss in the following a number of important issues that we have observed throughout our projects.

In a market driven health economy, where patients and insurance companies pay for the bulk of health care costs, or in situations, where research funds are available, it is to be expected that telemedicine solutions which increase quality of care or which reduce costs will be developed and marketed without direct stimulation from the health care authorities. Most successful telemedicine projects have been driven by either the needs of the involved clinicians or by a clear busines model based on economic considerations. Technology driven projects have been much less sustainable; and those projects organised by health authorities did often not meet the needs of the targeted clinicians well enough to be widely accepted.

In countries with limited resources the situation is fundamentally different. The majority of the population lives below the poverty line; the tiny upper-class that can afford private health care are often seeking it abroad. Thus, basic health care services for the masses are provided by the state and often by charity organisations. Even if telemedicine is proved to efficiently and effectively improve care delivery at the level of primary care, it is unlikely that it will be adopted by primary care staff on their own initiative as there are usually no means available. In a resource-constrained area telemedicine can only provide a sustainable and beneficial outcome for all if it can be implemented on the health system level. The challenging question for the future development of telemedicine is if and how it can be scaled up and integrated to a level where it can really strengthen a health system at large.

In the situation where resources are available for a bottom-up approach in which the health practitioners can organise themselves and choose how to use telemedicine, then the role of health authorities is mainly to develop appropriate guidelines and standards on security, safety and ethical issues of telemedicine. However, if a health care system is centrally organised and if the implementation of telemedicine is considered on a regional or national scale rather then by the individual (primary) care providers, a number of fundamental aspects must be considered. Successful implementation of telemedicine will depend on 1) beneficial outcome, 2) acceptance by all involved partners and 3) cost effectiveness for the health system[1].

15.1. Potential Benefits

Telemedicine is an exchange of knowledge and information between geographically separated partners of a health care systems with the aim of improving health care delivery. Information is exchanged between defined partners for a certain purpose. If we want to understand the benefits of such an information exchange we must understand the different purpose, of information exchange and we must consider its value for the different partners involved in a health care system. The less resources that are available the more important it is to implement telemedicine in a way that as many of its potential benefits can be unleashed.

15.1.1. Purpose of Information Exchange

The purpose of transferring information from one location to another can be that of a consultation, it can be education of health providers or it may also be general health information/knowledge access (fig.15.1). Consultations arise from an immediate medical problem – one individual within the health system (patient or clinical staff) approaches a distant expert with a concise question and the information returned is tailored especially to this very situation. Consultations are generally most closely associated to the term telemedicine. However, if we recall the definitions of telemedicine in the introduction (page 10), distance education and distance access to health information (evidence) are as much part of telemedicine – distance medical collaboration – as consultations.

Typical educational applications are driven by teachers utilising communication technologies to deliver educational material over a distance to the periphery (primary care, patients). While e-learning and tele-education are only slowly getting common reality, access to medical information has already become almost unthinkable without ICTs. Today, library catalogues, articles in medical journals and scientific databases are accessed primarily in electronic form.

Short development cycles and global penetration of ICTs make it tempting to develop electronic systems for information exchange tailored to a very specific purpose only. Compared

---

[1] A major complication with the latter is that the costs of health provision are bourne by multiple entities. Even if telemedicine does lower total costs, it is very well possible that costs or work load may increase for one specific partner involved. For example, if primary care centres use telemedicine for referral planning, this increases the work-load of the primary care centre. Costs are saved for the patient and for the referral hospital. It is thus iportant to consider costs for the total health system, especially if it is financed by the state. (c.f. section 15.3)

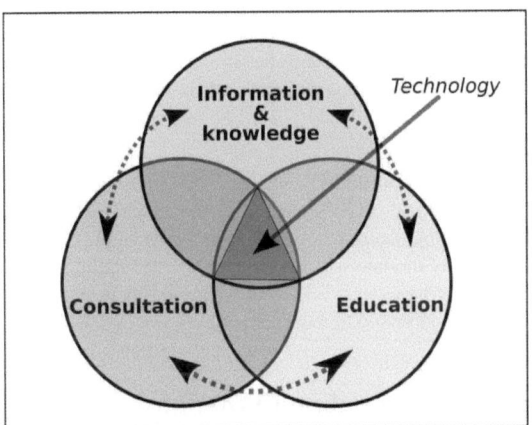

Figure 15.1.: Technology facilitates the exchange of information for different purposes. The large circles illustrate the main purpose of exchange in the medical field: access to medical information and knowledge (evidence base), education and consultations. Technology should place the users in the centre of information flow enabling them access to all kind of information. Interaction between the different fields are visualised with the dotted arrows (c.f. text).

to the traditional universal position of universities, there is in increasing tendency towards an "e-fragmentation". This neglects the fact that consultations/collaboration, education and research (resulting in generally valid information) cannot be separated from each other: A consultation often includes a fair amount of educative information; consultations provide valuable input on what type of continuous education is necessary; latest scientific results and evidence should always flow back into education; collaborations are important for quality research. In figure 15.1 these interactions are illustrated with the dotted arrows.

It should also be considered that technical access to health information (journals and books) is not by itself enough for successful dissemination of information. For many health providers, especially those who are not English native speakers, there is a high initial hurdle to use such information as evidence to improve their own practise. Initial human collaboration is often necessary to lower the barriers to start using available information. If such collaboration is not possible face to face, telemedicine offers a very viable option for such tele-tutoring. We have regularly observed that consultants would direct the case submitter to journal articles or even include the full text articles with their report.

For a person working in a functional academic environment this "e-fragmentation" does not impose any particular disadvantage, maybe except for the fact that an increasing number of different computerised systems and software have to be mastered. In areas where resources are constraint, where specialists are few and far between and where a general academic environment – providing access to information, education and exchange with colleagues (consultations) is not very well developed it is important to take care that technology is not increasing the gap between the partners of a health care system. For strengthening a health care system it is essential that all aspects of information exchange (consultation, education and knowledge access) are addressed.

Noteable, most of the feature request by users of iPath were in the direction of improving its capabilities of managing accumulated knowledge. This included the possibility to insert links to articles on medline and to attach full-text papers to consultations. Furthermore, a taxonomy system to precisely label any object within the database (images, cases, articles) is now in development. This taxonomy module will finally enable cross-linking of all content stored in an iPath-server. Another pending request is to create a possibility to re-use consultations as "learning problems" that can be presented to students.

15.1.2. Partners in Information Exchange

The primary aim of health care is to improve the well-being of the individual patient, and thus improvement of individual care is often perceived as a primary goal of telemedicine. However, taking direct impact of a telemedicine consultation on the outcome for the individual patient as sole measure would be falling short of the potential benefits of telemedicine. There are many other ways in which telemedicine, if properly implemented, can help to strengthen a health system and thus ultimately improve the individual's situation.

A proper understanding of the different possibilities how the health system can benefit from telemedicine is fundamental for a successful implementation on a larger scale. If the planning is focusing only on one type of application, it is much more likely that the achieved and perceived benefits will not outweigh the costs of the involved technology.

In this section I will go through different levels of a health system and summarise the potential benefits that telemedicine can offer at each level.

Patients

In the centre of interest in health care is usually the patient. Although the telemedical applications studied in this thesis are not patient-centric in the sense that the patient directly interacts with the telemedicine technology, there are still major benefits for the patients. The most commonly encountered benefit for patients is that telemedicine supports the local health provider with improved diagnosis which often translates into better treatment for the patient. A good example was presented as a case study in chapter 12. In this study telemedicine had a positive impact on treatment in over 50% of all cases.

Unfortunately it was not possible to study this impact on patients in the telepathology studies (chapter 9-11). Patient records were often not available and the capacities to conduct and document follow-ups were not available. In addition, for pathology diagnosis it is generally difficult to define clearly the outcome of a better diagnosis for the patient on the short term. Such effects are only measurable on a long term basis.

In places where telemedicine has been implemented on a regular basis in primary health care it could be observed that the mean duration of chief complaint could be drastically reduced[15]. The benefit for the patient was also reflected by the fact that "... all patients surveyed were either "very satisfied" or "satisfied" with their care, and most patients were willing to pay for a visit, with a median amount of USD 0.63"[15].

A second important impact on patients is that telemedicine can help to improve referral patterns. In rural areas the number of referrals from primary health care centres to district clinics is often minimal due to a low acceptance rate by patients. The opposite is true for urban areas where many unnecessary referrals lead to congested hospitals dealing with problems that could be addressed at a primary health care level[13]. In rural areas, distances between hospitals and primary health care centres are often long. Even if transportation is available, the cost of transport is a heavy burden for the often poor patients. Telemedicine can help to give the patient a better understanding about the necessity and implications of a referral.

However, the most important benefit for the patients will probably be a very indirect one. If telemedicine can help to improve the skills of local health providers (c.f. chapter 12) this will in the end make the largest difference for patients in general as the local health provider will gradually improve his diagnostic skills and will thus be able to treat more patients more adequately in the future.

Primary Care

In many parts of the developing world primary heath care centres, clinics, and small hospitals are the backbone of the health care system and deliver the bulk load of medical care. Often these centres do not even have fully trained medical doctors but are run by nurses or clinical officers. While nurses are skilled in direct patient care, they do normally not have adequate training and experience in diagnosing certain types of diseases. In a situation where the health professional at the primary health centre cannot diagnose the type of disease, the patient is typically referred to the district hospital.

Bossyns et al. [13, 14] observed that these referral systems are often functioning very poorly in rural Africa. Besides financial constraints, the noted obstacles are passive patients that do not understand properly the need for a referral and primary care staff that are "... reluctant to refer because they see little added value in referral and fear a loss of power and prestige"[14]. The result is poor communication between health facilities which again hampers the well-functioning of the health system[78].

In contrast to physical referrals from which the rural clinics usually do not get any feedback telemedicine poses the primary health worker in the center of activities. Tele-consultation with the specialists from the referral centre can either provide the primary health staff the necessary diagnostic input to treat the patient locally or they may confirm the necessity physically referring a patient. In addition, tele-consultation would provide the primary health centre with an implicit continuous medical education which is often very rare if not totally absent in rural Africa.

In many developing countries young doctors are obliged to do at least part of their residency in form of "community service". They are often reluctant to do so because they do not yet have the clinical experience in dealing with a lot of the medical problems encountered. Furthermore, in order to foster their own personal career and in order to qualify for full registration with the medical board they need to get specialist training. Traditionally, the only way to get advanced training is working in a larger medical centre where senior specialists are available. A telemedicine link to appropriate specialists could improve the situation for doctors on community service by giving give them the possibility to stay in continuous contact with their tutors for clinical consultations as well as for continuous training[53]. In a needs assessment carried out with medical residents in Cameroon, Scott et al [134] found that most of the surveyed residents (n=17) indicated that the ability to contact a mentor would have altered their handling of recent cases (84%). Every resident had access to a mobile phone and 65% had used it to contact a medical colleague for guidance.

It is noteworthy that the benefits of telemedicine for primary care in developing countries has hardly been demonstrated. Benefits, especially for better referral planning, are only possible if telemedicine is well integrated into the health system. This is a big problem in developing countries. Advocates of telemedicine prefer isolated high tech solutions over integrated services that rely on basic technology like telephones, VHF radio, etc. Equipment from the industrialised world is often not directly applicable, and technology that is appropriate is often not considered, partly due to lack of local specialists.

Medical Specialists

For medical specialists there are different types of potential benefits. Especially in developing countries and rural areas there are few specialists and thus they often work under professionally isolated conditions. Outside the main medical centres opportunities for continuous medical education are rare. Telemedicine is very suitable for overcoming professional isolation and it can give access to quality assurance and to continuous medical education without expensive and time-consuming travelling. Telemedicine offers an opportunity to get a second opinion on a complex case even if there is no other specialist available in the same place.

Medical sciences are progressing very quickly. Diagnostic tools as well as therapeutic interventions are changing and the life of a specialist is that of constant learning. Medical professionals in developing countries are often cut off from new developments due to various reasons: access to scientific publishing and updated text books is limited, exchange with colleagues is hindered by distance, coordinated continuous medical education activities are non-existent and travelling to international conferences is often not affordable. For isolated specialists modern ICTs offer ways of communicating with other specialists. The histopathology forum (c.f. page 58) is a very good example of this type of application of telemedicine. The forum is organised as a virtual community with participating pathologists from all continents. Consultations and case presentations have so far been posted by 18 different pathologists from India, Brazil, Lithuania, Armenia, United Arab Emirates, Myanmar, Uzbekistan and from other countries. Over 50 pathologists have participated in discussing these cases by providing not only diagnostic input but also literature references and occasionally copies of full-text articles. The log files of the server reveal that there are many participants in the forum that do not post their own cases or write their own comments, but who are regularly following the discussions. This clearly indicates that for many participants following a case discussion of senior pathologists is at least an educative past-time if not some form of unorganised CME. Particularly interesting is the recently increasing sharing of literature references on iPath. Only some pathologists have good access to scientific literature, either through their library or through programmes like HINARI[2]. In addition, the huge amount of available literature makes it difficult to find good articles which are relevant to a specific problem, epscially for junior specialists.

In many medical fields clinical decision-making is based on the diagnostic input from a number of specialists. Oncological treatment plans are typically based on the information provided by pathologists and radiologists. Difficult cases are frequently discussed in "tumour board meetings", which bring together radiologists, pathologists, oncologists and other clinicians involved in the treatment of the cancer patient. Such *"virtual care teams"* can be greatly facilitated with the help of telemedicine. There is currently no example for such an application from a resource-constrained area, but the "Oncology Centre Lörrach" (c.f. page 56) may serve as illustration. Since 2002 almost 500 cases presented and discussed at the institutionalised oncology meetings in Lörrach have been documented on iPath jointly by the oncologists, pathologists and radiologists, who are notably all working in different locations. Today the telemedicine tool has become an integral part of the oncology meetings: interested team members can look up the material before the meeting; during meetings all material is presented directly from the telemedicine system and no longer from transparencies, slides and PowerPoint presentations; conclusions are often annotated in the system and after the meetings every participant can at any time from anywhere review all material. As a side effect, the group has built up database with over 400 oncology cases containing relevant radiology and histopathology images which now build a reference and teaching repository.

At present the medical specialists are certainly the sector of the health system in developing countries that can most readily benefit from telemedicine. Often they already have or can afford basic infrascruture: essentially this is a digital camera and a PC with an Internet connection. Their medical background also allows them to relatively easily integrate with international communities of specialists: through communication by email or by participating in on-line discussions such as those on iPath or by contacting specialised consultation services such as AFIP telepathology[163], the UICC telepathology consultation centre[40, 101] or telederm.org[136]. It is worthwhile noting here that collaboration with developing countries can also be interesting for specialists in the industrialised part of the world. Several pathologists working as consultants on the iPath server at the University of Basel have indicated that a motivation for doing so is, apart from the humanistic aspect, the fact that they get to see many interesting and rare pathology cases which are hardly seen in Western Europe.

Health Care System

The impact of telemedicine on a health system is very difficult to assess. Improved diagnosis for individual patients and skills development for the health professionals are only one part of the potential benefit of telemedicine for health care systems. In many African countries the health system is organised by a two-tier district approach with health centres offering proximity care with relatively low technology while district hospitals provide back-up for patients referred by health centres[14]. Although proposed referrals are generally accepted by the patient, the compliance rates are often relatively low due to various reasons. Bossyns et al. report that in rural Niger "62% of emergency referral proposals for children below 5 years were not complied with"[13]. Even if patients visit a district hospital, there is usually no feedback from the district hospital back to the nurse at the health centre. These circumstances are neither a satisfactory solution for the patients nor very motivating for the health workers to improve their own skills and abilities as they perceive the situation purely as lack of means and resources available at the primary health care level[14]. Telemedicine would be an ideal tool to address both issues. A telemedical consultation could clarify if a physical referral of the patient is really necessary, and the consultants could also provide some information about the expected duration and planned interventions. Many referrals could become unnecessary as the treatment could be delegated to the health centre. As the health centre is initiating the referral process with a tele-consultation, there is an immediate feedback for the nurses. Even in the situation of a physical referral the physicians at the district hospitals could even use the telemedicine link to provide the health centre with records and to receive follow-up information on patients' status after they returned back to their homes. An illustration of the capabilities of ICTs in terms of preventing unnecessary referrals was provided in fig.12.3.

Telemedicine can also play an important role in establishing disease registries or generally in collecting epidemiological data. A problem with classical reporting systems is that they are often organised only in one direction where the periphery reports towards the central authorities. Health staff are sometimes reluctant to provide accurate and timely data and statistics as they do not get any immediate feedback or direct benefit from their reporting. Although telemedicine cannot change this problem itself, the bi-derectional communication provides an immediate feedback. Furthermore, consultative activities can give also a certain immediate picture about some of the health problems within a region.

15.2. Acceptance

Even if theoretically benefits of telemedicine are convincing this does not imply that telemedicine will find good acceptance with all stakeholders involved. Health care professionals are working on a very busy schedule and there is not much time to learn about new technologies.

In many developing countries nurses and doctors get no or only minimal exposure to ICTs during their training. Similarly, many of the senior medical specialists that are willing to participate in telemedicine projects do not have an interest primarily in technology, but in sharing their knowledge in a particular field of medicine. In our experience there are two technologies that most health professionals master on their own – mobile phones and Internet/email. Chances that telemedicine is accepted by the involved medical staff are higher if it is based on familiar technologies. In contrast, if staff have to deal with too much special equipment which does not have any practical application within their normal work but suits telemedicine only there is a higher chance that technology will not find good acceptance. The experience with video conferencing in South Africa may serve as an example. In some provinces a number of video conferencing systems have been set up within a national telemedicine project. Especially in rural settings several problems could be observed. Many professionals were reluctant to learn how to use the equipment on their own as it was perceived as something that they knew they would never use for their purpose. The same people all had mobile phones or TV sets and mastered to handle those easily without any particular training effort.

A fundamental misunderstanding is that acceptance will increase proportionally with interactivity and speed of connectivity of a telemedicine link. The medical professionals perception of "speed" of a telemedicine system is not directly related to its true technical speed, but rather on how quickly they can finish a task. Synchronous telemedicine solutions such as video conferencing (VC) are able to accommodate immediate tele-consultations, but only if all partners are available at the same time. This requires a tight schedule and it implies that a delay at any side will equally affect all participants. In the busy day of a health care professional any unnecessary delay is usually perceived very negatively. In situations where store-and-forward is possible it is often much more feasible and acceptable although it does not offer the interactivity and speed of real-time consultations. A nurse in a primary care centre can easily take images of a dermatological lesion with a digital camera and send them with a short clinical description to a dermatology consultant. The consultant can review the images later independently. Reviewing a few still images and writing a short report is often much quicker than waiting for the patient to undress and pose in front of a VC-system and then to explain to the patient and nurse the findings and further proceedings.

15.3. Cost Effectiveness

A common problem with the implementation of any new technology is its legitimation with regards to cost effectiveness. For technologies such as telemedicine which involve partners from different levels of a health care system it is very difficult to establish actual costs and savings[63, 160]. Direct investments for infrastructure vary greatly depending on the chosen technology. Many real-time telemedicine applications call for highly specific and expensive technologies such as video conferencing or robotic microscopes and dedicated high speed communication lines. In contrast, store-and-forward solutions often do not require anything except a PC, which may already be present in many cases, and a simple digital camera. If telemedicine can run on existing hardware and communication lines, the additional direct costs are minimal.

However, looking at the hidden costs gives a much more complex picture because the application of telemedicine often changes the medical work-flow drastically and shifts costs from one entity to another. While the application of telemedicine may save costs at one side there are increased costs somewhere else. In addition, the main aim of telemedicine is often an improvement of diagnostic security, reduction of turn-around times or the educational value of information exchange. All these factors do not directly translate into cost savings. Estimates about cost effectiveness of telemedicine can thus not be given in a general way, but must always be determined with respect to a concrete application. If the costs for travelling or visiting a specialty clinic are relatively low, telemedicine may not be cost-effective. However, if the benefits of technology and knowledge transfer are considered[75] or if transportation is expensive[12, 87, 88], investment in technology may be cost-effective.

15.4. Deployment Strategies

Advocates of telemedicine have raised hope that this technology will solve many of the problems of delivering quality care regardless geographical distance. It is thus tempting to believe that the situation in rural clinics will improve automatically with the installation of the necessary technology. The experiences from national/provincial telehealth projects, e.g. from Australia[33, 180] or South Africa[60], teach us that it is not as easy as installing some nifty technology. Alhough installed equipment was technically working it was often severely underutilised. In a country like Australia, where basic infrastructure such as telecommunication lines and power supply is reliably available, it was found that it is not so much technology but the human factor that tends to determine success or failure of telemedicine. In resource-constrained areas this will not be different but the unreliability of basic infrastructure requires a much more careful selection of appropriate technology.

For the successful implementation of telemedicine on a larger scale it is crucial to consider a number of issues during the planning phase. In order to be successful, telemedicine projects must be technically feasible, medically valid, reimbursable or government-funded and they must be institutionally supported. Although the projects analysed in the scope of this thesis have all been of a relatively small scale, it is possible to learn some lessons from the limited experience and to draw some conclusions for the future.

15.4.1. Selection of Technology

Selection of appropriate technology is of crucial importance. Investments in technology are often the highest initial costs of telemedicien projects, especially in areas where the penetration of technology is generally low. Key issues for selecting technology include the following:

Considering the Weakest Link

When planning a telemedicine network it is essential to consider the (technologically) weakest link. The fanciest telemedicine receiving station is useless if the sending station is not operational due to lack of stable electrical power or due to unreliable telecommunication lines. The highest image quality is useless if transmission lasts hours. If collaboration with partners in low resource settings is envisaged, there must be at least a fall-back method that works relatively reliable also for the most remote site of all partners involved. When we started the collaboration with the National Referral Hospital on Solomon Islands (c.f. chapter 9 & 10), iPath was only offering web-based access. The experience from submitting the first few cases over an extremely slow connection and the frustration when transmission was stopped in the middle brought up the idea to use email to submit cases. When submitting a case by email, the physicians in Honiara could prepare all cases off-line in their email client (Outlook Express) and then send them all at once in the background telling the email client to disconnect the modem as soon as all emails were sent. Since the common email protocol (SMTP) is very robust, we hardly observed any loss of data even on the slowest network. Another advantage of communication by email is even much more profane. Before implementing the email interface for iPath, physicians had to log in on the web site to check for diagnosis on their latest case. After the introduction of email they could receive all comments conveniently in their personal email which they were reading regardless of the telemedicine activity. After some time it happened that certain case submitters had forgotten that they were actually communicating by email with a web server and not with a person.

Using Common Technology

In order to keep initial installation costs at a minimum but also in the interest of long-term sustainability it is useful to deploy commonly available technology. A frequent mistake in telemedicine implementations is not to plan for continuous support of technology in rural areas. It is much easier to find on-site support (commercial or on voluntary basis) for common technology such as dial-up Internet connection over phone lines and ordinary PCs with standard software (email client or web browser) than it is for some special purpose hardware. A pathologist in Bangladesh was waiting for months to get the CD-Rom with the software for the special purpose microscope camera that he had purchased because the companies sales representative would visit Bangladesh only few times a year[2]. The software for more common consumer-type digital photocameras can be usually downloaded from the web and the local vendors can often also organise copies of the software.

If telemedicine technology is not regularly used, there is also the danger that a minor fault developing in-between two sessions is not detected and that subsequently the system is not readily operational by the time when it is needed. For example, video conferencing systems usually rely on ISDN-lines. In primary care facilities in rural South Africa such lines are often installed exclusively for this purpose. If there is a problem with the ISDN-lines, it will stay undetected until the next video conference session. A problem on the ordinary phone lines would have been identified by the first person trying to make a phone call.

Another issue is that the use of common technology usually requires less specific training. Health

---

[2]The necessary software was finally organised from the representatives in Switzerland and was sent to Bangladesh via Basel.

care professionals have a reputation of being extremely reluctant to use new technologies. If telemedicine relies on some special purpose technology that health staff would not use otherwise, it often requires quite some motivation to win their interest and they will need extra training. If it is possible to use a more common technology that they are already familiar with – for example mobile phones or sometimes Internet/email – the hurdle to use it also for telemedicine is less high.

Multi-Purpose Technology

The basic infrastructure for telemedicine is often not specific and can be used for different purposes. Costs could be minimised by sharing infrastructure between different applications of ICTs. This effect is not necessarily obvious in pilot projects, where telemedicine is often regarded as a totally separate application: a personal computer with an Internet connection may be even more costly than a video conference station. However, if looking at costs from the perspective of the health system, then a PC can potentially serve many more purposes than the VC. The PC can be utilised for general office work, email communication, literature searches, record keeping and statistics as well as for telemedicine.

The feasibility of using general purpose equipment for multiple purposes largely depends on the kind of medical problem that is being addressed by telemedicine. In a mission critical situation, where the patient's life is at stake and where remote interventions or diagnoses are required, the reliable and safe operation of telemedicine is of foremost concern. The situation in most developing countries, however, is different insofar that the prime goal of telemedicine is often to support and empower the rural health care providers. Besides, the application of one type of telemedicine is often inducing the need for another type. If pathology diagnoses are becoming readily available, the next question will be how to proceed with treatment. Thus, taking into account the fact that there are a number of different applications of telemedicine (c.f. section 15.1), it is advisable to plan right from the beginning towards a network allowing different kind of applications if possible based on one common technological platform.

15.4.2. The Process of Implementation

Probably the most important single factor for a successful use of telemedicine is the process of implementation. The implications of integrating telemedicine into health care systems and the associated problems have been extensively studied in Canada by Jennet et al[71, 73, 72]. They found that the integration of telehealth into health systems benefitted from staged implementation and from building on each community's cultural context. Keeping records of best practices and lessons learnt as well as systematic evaluation are significant for the roll-out and success of future implementations. They also stress the importance of 'readiness' of the environment which can be improved by means of a staged approach that gives health staff, administrator and the communities time to adjust to the changes associated with telemedicine.

The most 'telemedicine-ready' community in a health care system in developing countries are the medical specialists. They are aware of the importance of continuous medical education and interaction with colleagues and are thus very susceptible for the possibilities of telemedicine. In primary care, especially in rural areas where the exposure to communication and information technology is minimal, the situation is quite different. It will need concise strategies to create awareness on different possible applications of telemedicine technology and to minimize the barriers that often prevent their use. In the following I will list the most important strategies developed or observed throughout our own projects.

The Local Clinical Champion

An important role in the implementation process is that of the local "clinical champion"[47, 145]. In the initial phase the clinical champion's role is that of a visionary pioneer who can create awareness and interest. Later the role changes to that of the advocate and coach. The clinical champion has an important role in lowering the behavioural, technical and institutional barriers that often exist towards the use of telemedicine. Ideally, the clinical champion is a doctor or a nurse working in a clinical setting: this greatly helps to keep telemedicine focused on the clinical needs.

Personal Motivation

Health care professionals are frequently burdened with an immense workload and tight schedules. Most clinics in developing countries are not adequately staffed to deal with the amount of patients and there is often additional administrative work that is left with the health care professional due to a lack of administrative personal and an inefficient administrative organisation. Most primary health care staff in developing countries are not familiar with computers and other information technology and there is a tendency that new technologies such as telemedicine are perceived only as additional workload. Thus, there is generally a high initial hurdle to use telemedicine and the idea is not often encountered with much verve except with the few enthusiasts. In this situation it is often necessary to communicate the potential benefits of ICTS and to create a certain motivation for people to start using it.

Motivation is generally stimulated through the appreciation and acknowledgement of a person's work. If an individual feels that his or her work is having a positive impact on his or her environment, this person is more motivated to improve his or her own skills. On the contrary, if a someone feels that one's efforts are not being recognized and that there is no appreciation of what one is doing, motivation will decrease. It is thus important to convincedemonstrate to primary health staff how telemedicine can help them improve their own skills and thus achieve a better outcome within their community rather than having to refer patients to the next level health care facility. In addition, usage of telemedicine will increase the "visibility" of primary care staff, it will enable them to get a feedback from the next level health facility and finally, if patients can be treated successfully within the community, increase the health workers' appreciation within their community. If telemedicine is utilised in appropriate manner it may also serve as a motivating factor.

Personal Interaction

Motivation is often based on personal perception and on interaction between individuals. In telemedicine, where collaborating individuals are normally geographically separated, it is important not to neglect the component of personal interaction. Especially in the initial phase when technology is new for its users, it is valuable to give them some feedback. As an example, in the telepathology project on Solomon Islands it has been very helpful to urge the surgeons to pass on the pathologists' feedback on slide quality to the laboratory technicians. Knowing that their slides were reviewed by international specialists and knowing that the slide quality was appreciated was a real motivation for the technicians to keep improving the quality of histological slides. Although the laboratory is technically primitive, the quality of slides produced today is comparable to those produced in Switzerland.

We found it very useful to provide some sort of initial tutoring. If a new person wanted to start telemedical consultations through our system, we often tried to open two channels of communication: consultations and mentor. The telemedicine consultations are directed to the iPath telemedicine platform where the role of the consultant is not tied to an individual person, making it much easier to triage consultations to the optimal specialists. Since the consultants are working in a group, they have some freedom of who should answer which consultation and we could observe that there was often communication between the consultants, too. However, besides the consultations, we often tried to open a personal communication between the new non-expert and one of the consultants. This private mentoring channels have been useful for communicating certain guidelines and also some criticism on how to improve the quality of submission (image quality, clinical information, etc.).

Strengthening of Existing Partnerships

From the many telemedicine collaborations that started using our iPath server at the University of Basel we have observed that the most successful ones were those that built on an existing international partnership. Once telemedicine has been implemented successfully in the scope of an existing collaboration it is much easier to introduce new applications and to extend the usage of the telemedicine link to areas outside the initial collaboration than it is to newly introduce collaboration by telemedicine only.

Evolving rather than Imposing Guidelines

The accuracy of telemedical consultations depends largely on the clinical data (patient history, complaints, images, etc.) recorded by the non-expert. Most non-experts have neither experience with selecting and photographing lesions nor with the written recording of a clinical history.

While it is certainly necessary to develop certain standards for recording clinical data, it is is also a fact that complex forms for data capturing or very strict guidelines impose high barriers for the usage of telemedicine, especially for people with low previous exposure to technology. It is very helpful to *guide* new users of telemedicine towards certain standards rather then trying to enforce them.

A Strategy for Procuring Technology

Beside the selection of technology itself, the process of procuring technology can be a key determinant for the success or failure of telemedicine. Pilot projects are often implemented using donated or specifically purchased equipment. While in some situations the application of specific technology may be justified, telemedicine can generally be implement most economically using common technology which can be easily shared with other applications. While sharing computers and connectivity between different tasks, i.e. utilising infrastructure for multiple purposes, is economically sensible, it requires careful planning to prevent that equipment ends up monopolised for one use or even largely unused because it is not accessible to those who should use it.

The Role of Health Administrations

While he most prominent role of health administrations is that of funding technology it cannot be stressed enough that the selection of appropriate technology should be based on clinical needs and not on administrative reasoning. Health administrations should probably not select technology but rather develop the framework in which technology is to be deployed. This includes the development of policies but in our experience even more important is the organisation of workflow and communication between rural health providers and medical specialists. For example, the availability of mobile phones alone does not solve all problems. Even if all residents and nurses are equiped with mobile phones they must still know whom to call for what questions. On a large scale, such a "telephonic referral system" will not work if it is based on a personal contacts only. An unavoidable step for a large-scale deployment is a proper organisation in form of a call centre or a store-and-forward system.

15.4.3. Decentralised Collaborative Networks

Potential benefits of telemedicine for a health care system are diverse. There will not be one single technology that suffices all these different types of applications, but for a successful implementation at a larger scale it is important to develop a strategy that encompasses all aspects of electronic information and knowledge exchange (consultation, education and access to medical information).

Many telemedicine projects are essentially outsourcing certain medical or educational services to foreign institutions. There are many institutions offering services at a distance, some of them for free and some of them on a commercial basis. For instance, the British Medical Journal offers excellent on-line learning modules for general practitioners[3]. Outsourcing services is certainly the solution that requires least organisational effort. However, it bears several problems. Outsourcing creates external dependencies – for content as well as for technology. Usually the service providers will determine the technological requirements for accessing their services and often it may not be easily possible to supply the necessary infrastructure. At the content level the situation might be even more problematic. It was found in a pilot project between West Africa and Switzerland[53] that "... several topics for seminars, requested by physicians in West Africa, could not be satisfactorily addressed by experts in Switzerland, due to major differences in diagnostic and therapeutic resources and due to discrepancies in the cultural and social aspects".

When addressing the primary care level it is crucial to include the regional medical specialists and to rely on technological infrastructure that is commonly available and not one that is requested by service providers alone. A promising way to achieve this is fostering South-South exchanges of expertise through decentralised and regional collaborative networks. Timely, secure and reliable exchange of sensitive information for the benefit of primary care providers requires a good organisation of the collaboration between all stakeholders of a telemedicine application, particularly for store-and-forward telemedicine where communication is not synchronous. The experiences with telepathology on Solomon Islands and Cambodia have led us to the development of computer-aided virtual communities of experts who jointly provide services (virtual institute). On an international level such services organised across institutions and borders raise legal, ethical and economical question that have not been adequately addressed to so far. However, on a regional level the legal and ethical situation is much less problematic. Joint expert services could be implemented based on existing forms of collaboration, beneficially supplementing the existing referral system. While such services are ideally organised by regional authorities and specialists, this should not prevent participation of international specialists in such a regional network. For the medical specialists the possibilities to collaborate at an international level are often more interesting than collaboration limited to a regional level only[25].

Implementation of telemedicine in form of decentralised networks imposes far greater challenges than participation in an established international network. While installation and maintenance of technology and training of the users in its application is a challenge of its own, the organisation of work-flow and collaborative frameworks are much more important. Successful and sustainable implementation of telemedicine lie in the social processes of human resource development, changing organisational collaboration and the dissemination of relevant information. It is import to realise that technology does not only serve the purpose of transmitting medical data from one point to another. For the organisation of a larger network of collaborating specialists it is indispensable to automate the flow of communication to some extent by deployment of computer-mediated communication and social networking approaches.

---

[3] http://www.bmjlearning.com/

Bottom-up Approach versus Central Organisation

For implementing telemedicine within decentralised regional networks in resource-constrained areas there are two somewhat contradictory principles that have to be balanced. On the one hand it has been repeatedly reported that telemedicine is most successfully implemented when driven by local clinical champions based on clear clinical needs and organised from the bottom-up. On the other hand, implementation of telemedicine as a tool for strengthening a health system at large needs some central coordination, if not only the few hospitals and GPs who already possess the necessary technical skills and connections are to benefit. Many of the potential benefits of telemedicine, such as reduction of unnecessary referrals and skills development of primary care staff, can only be economically viable at a system level since they are not equally cost-efficient for all partners involved. If the specialists' income is lower for telemedical consultations than for physical examinations, they will usually prefer physical referral of patients although this imposes much higher costs for transportation, bourne by the patient usually, and provides no benefit for the referring primary care centre.

Technological infrastructure necessary for telemedical activities is not available in many clinics. However, basic ICT infrastructure can serve multiple purposes such as patient information management, planning (e.g. through a district health information systems[164]), access to health information, knowledge management as well as tele-consultations and distance education – in the medical fields but as well in other sectors. Thus, wherever possible deployment of technology should be preferably organised in a multi-sectoral approach, e.g. in the form of general purpose Internet access points, enabling not only telemedicine but benefiting a wide variety of applications in education, local economy, etc[53]. The project in the Tsilitwa clinic (chapter 12) provides an example of such a multi-sectorial ICT deployment and for the Solomon Islands Telemedicine Network (c.f. 16.3) a similar approach is planned by fostering a collaboration between a UNDP project for rural connectivity and the Ministry of Health.

Open-Source Approach

Open-source software is becoming increasingly popular. In the medical field there are several open-source projects developing software for patient and practice management. Open-source offers two main benefits which are important for resource-constrained areas. Most open-source software is available without licensing costs and can be freely redistributed. Thus, large-scale deployment and software upgrading do not involve high licensing costs. More important, however, is the fact that open-source software is open for adaptation and further development. Software engineers from developing countries can easily adapt open-source software to local needs. A prime example is the care2x project[4], which provides an open-source hospital information system including patient management, billing, laboratory and pharmacy management and much more. Care2x is used e.g. in Tanzania by a number of hospital and it has been adapted to local

---
[4] http://www.care2x.org/

languages and work-flow. In addition, a national project for exchanging data between care2x and DHIS[164], notably now also an open-source project, has been initiated[5].

Despite clear advantages, we have experienced that open-source is often perceived by health administrations as unprofessional, inferior and insecure. The arguments are usually not based on any technical evaluation but rather derived from the fact that open-source cannot be bought from a "reputable" company that will guarantee proper functioning and maintenance. We observe a tendency that health administrations prefer to buy telemedicine in form of fixed services and packages. It seems to be a problem with open-source software that it cannot be bought from a vendor as complete solution.

While there are no licensing costs with open-source there is a need for skilled technicians who are able to install and maintain open-source solutions. When promoting the implementation of telemedicine or other IT solutions in health based on open source it must be also be considered how open-source solutions are going to be supported[6]. While we tend to see it as an advantage that open-source solutions leave more resources for development of local human capacity, decision-makers often prefer spendnig on hardware over spending on human resources as hardware seems to present a more solid value.

15.5. Towards a Grid of Decentralised Networks

While an approach of decentralised regional networks seems very promising for fostering regional collaboration and better integration with the regional health system and for addressing technical implementation and support, there are also activities that are favourably happening at an international level. In addition, in many developing countries health care systems are often substantially supported by international organisations. Even within the limited scope of this thesis it has become obvious that the co-existence of regional organisation and global information exchange is a key factor for successful deployment in resource-constrained areas. In our projects this was accommodated by the possibilities for collaboration between the various virtual communities hosted on our server at the University of Basel.

For a larger scale implementation within health care systems such networks must be technically and organisationally independent. While patient consultations and referral planning should be preferentially addressed at a local level, the need of regional specialists to further consult with colleagues on an international level should also be accomodated. Relevant educational content produced at an international level should be easily accessible through regional networks extending into primary care.

---

[5] Some information on care2x in Tanzania is available from these two websites:
http://www.hisptanzania.org
http://health.elct.or.tz/projects/Care2x.htm

[6] It should be noted that similar problems can also be observed with many commercial products as user support and services in developing countries is often minimal.

There is a huge potential for inter-linking such different networks, at present, however, there is little understanding about how such independent networks can be integrated into a loose grid of networks allowing transparent information exchange and inter-connectivity at all levels. Grid technology is an emerging field of computer sciences addressing several issues in distributed computing and providing concepts and frameworks for the technical integration of networks[80]. Nevertheless integration of different telemedicine networks into a grid is not depending on technical standards alone – organisational, psychological, social, financial, legal and political aspects must be equally addressed. This will be one of the challenging areas for further research in telemedicine which will be of high impact also for its application in developing countries.

15.6. Recommendations

The experiences gathered and lessons learned lead us to a number of recommendations which we will try to incorporate in future projects.

- The organisation of work-flow and collaboration in telemedicine should be driven from the clinical needs. The doctors must be involved in the process.
- Support clinical champions and empower the health staff to benefit from ICTs in their own best way by providing infrastructure and training but without imposing total solutions. Applications must be adaptable to the local care process.
- Foster multi-sectoral application of ICTs by promoting telemedicine tools that are based on technology that can be shared with other applications. Evaluate and develop models how to utilise commonly available technologies such as mobile phone in telemedicine networks.
- Promote pilot projects that focus on applications that are replicable at a health system level and study the possibilities for integration into regional networks.
- Create and publish databases of pilot projects (successful and failed) and best practises and evaluate projects in terms of impact on individual care, involved care provider and health system.

16. Ongoing Projects

16.1. Educational Impact of Telepathology Consultations

In a first study we analysed the quality of telepathology consultations. Besides the delivery of an accurate diagnosis for the patient such consultations provide an important input for the local pathologist to develop skills and to achieve a certain security for making diagnoses locally. Since the beginning of the study in 2002 the pathologist in Phnom Penh, Dr. Vathana, has made his own diagnosis on any case prior to submitting it via telepathology. The planned study we try to analyse the development of the quality of these diagnoses. If the results are positive it will indicate that telemedicine provides an efficient tool for knowledge transfer and that on a long-term basis the role of telemedicine may change from delivering primary dianogis towards providing quality control.

16.2. Telecytology Study at NRH

During the telepathology project in Honiara the demand for diagnoses in cytology has become more and more prominent. However, compared with tele-histology, there are very few documented experiences in tele-cytology. In order to evaluate the feasibility of telemedicine in cytology the following controlled clinical study has been designed.

Control Study in Basel: A cytologist in Basel (L.Bubendorf) captures images from randomly selected routine cytology slides processed in the cytology lab at the Department of Pathology in Basel. These images are placed on the iPath platform and evaluated independently by four specialists (P.Dalquen, K.D.Kunze, P.Spieler, H.Neumann). These remote diagnoses will be compared with each other as well as with the original diagnosis of the cytologist in Basel (gold standard). So far, 170 specimen have been collected and remotely diagnosed independently by the four specialists.

Cytology on Honiara: The laboratory in Honiara has started to produce cytology slides from fine needle aspirates and smears (gynae, sputum). These slides are processed and stained according to a similar protocol as used in Basel. The slides are screened by the technicians/physicians in Honiara and they capture images of suspicious cells and place these images on the iPath server at the University of Basel. The same collective of specialists as above will review the images and make their diagnosis. Each pathologist will independently review all slides and images in order to exclude intrer-obrsever differences.

During the duration study, the original slides are sent by air mail to the Queensland Pathology lab in Brisbane, where they are reviewed by experienced cytologists. The diagnoses of the specialists in Brisbane will be regraded as gold standard and will be compared with the tele-diagnosis. If possible the slides will also be retrospectively reviewed by at least one pathologist from Europe.

Expected Results: The prospective design of the study will allow us to establish the level of quality possible and also to identify the weak parts in the process in order to identify what training would be necessary and most effective to improve the quality.

16.3. Solomons Islands Telemedicine Network

Solomon Islands are a group of roughly one thousand islands, scattered over hundreds of kilometres in the South-West Pacific Ocean. Distances between islands are long and many islands are very rugged and overland transport is extremely limited. These seem to be the typical preconditions for the application of telemedicine. While the telepathology project[1] had been running for years and enabled over 450 international consultations between the National Referral Hospital (NRH) in Honiara and the University of Basel, there has been little use of telemedicine *within* Solomon Islands. The Swinfen Trust has installed telemedicine links in some provincial hospitals[106, 142] which allow consultations with international specialists. Additionally, there have been occasional consultations by email from provincial hospitals to the specialists at NRH. However, with changing staff and a general shortage of doctors, tele-consultations based on personal email were not sustainable on a long term.

In 2005 we were approached by the doctors of NRH with the idea of extending the network that was established for telepathology to a general purpose telemedicine network available to the whole health care system. At the same time specialists from the St. Vincents Hospital in Sydney, Australia, were looking for ways of facilitating patient referrals from Honiara to Sydney. The two countries have an agreement that a limited number of severely ill patients from Solomon Islands can get treatment in Australia. The specialists from St. Vincent whished to have better control over the process of selecting patients that are eligible for the programme. In the past there had been patients referred whos medical condition was too advanced for receiving any treatment beyond palliative care; a situation equally observed in other projects[84]. It is now envisaged to use telemedicine to achieve a better selection of patients eligible for treatment in Australia.

Based on these two ideas a project for a Solomon Islands National Telemedicine Network is being prepared. This network should include the following components:

---

[1] chapters 9 and 10

1. Consultations between provincial health centres and NRH: submission of consultation must be based on email as this is the only technology available. The People First Network[2], a UNDP-project enabling affordable and sustainable rural connectivity and facilitating information exchange between stakeholders and communities across the Solomon Islands is operating a network of rural email stations, which could be used by the health care staff.

2. Triage: if possible, consultations should be answered by the specialists at NRH. If no appropriate consultant is available locally, an option to forward consultations to international consultants must be previewed. All consultations should be triaged through NRH-staff to ensure relevancy of replies for the rural clinics and to enable an educational benefit for NRH.

3. Referral planning: It is planned to utilise the telemedicine network as decision support tool for selecting patients eligible for treatment at the St. Vincents hopistal in Sydney. The network will enable different specialists to jointly review referral suggestions and to select the most appropriate candidates.

4. Continuous Medical Education: In collaboration with People First Network's project on Distance Learning Centres, distance education for rural health care workers should be implemented.

The major challenge of this project will be to place the NRH at the centre of the organisation and to interlink all these activities with each other. In order to support the staff at NRH with consultations from provincial health centres, it is previewed to collaborate with the specialists from St. Vincents and other international specialists. In addition, the physicians at NRH would like to use the telemedicine network to support and improve planning of patient referrals from provincial health centres. The project will create a framework in which the stakeholders of the Solomon Islands health care system will have a chance to learn and experience how the health system could benefit from available ICTs.

Australia is supporting health systems in many of the small pacific countries through its AUSAID programme and has several peace keeping forces stationed in the region (East Timor, Solomons). Many private organisations in Australia are organising medical support for neighbouring countries through various ways. If successful, the Solomon Islands Telemedicine project may well serve as a model how to utilise ICTs in order to improve collaboration between developing countries in the South Pacific region and Australia. Ideally, an international network of medical specialists could be shared between project in different countries and educational material prepared for one country could be more easily made available to other countries.

---

[2]http://www.peoplefirst.net.sb/

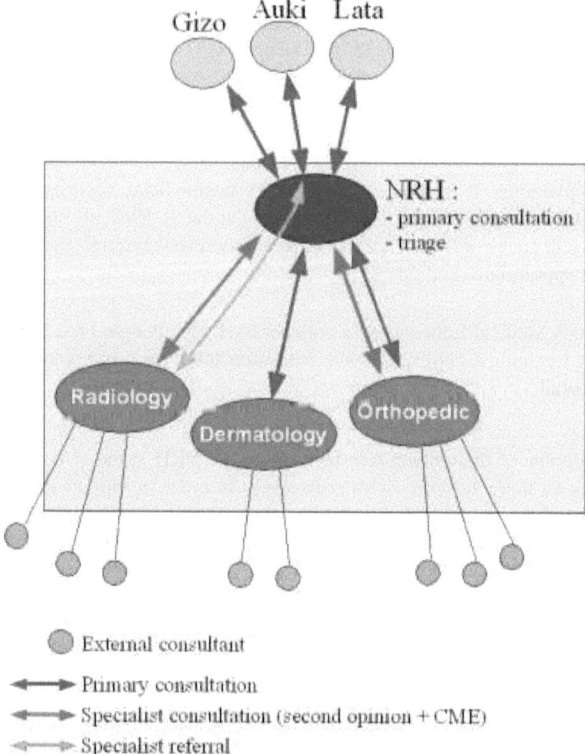

Figure 16.1.: Organisation of the Solomon Islands Telemedicine Network. At the centre of the network is the National Referral Hospital on Honiara. All consultations from provincial health centres are triaged by the staff at NRH. If possible, NRH staff will be answering the consultations. Expert groups in different medical fields are previewed and will include international specialists to support their patterns at NRH.

Current State

A project proposal for submission to funding agencies is currently being prepared by the Department of Health and the National Referral Hospital. The Australian NGO MedTechOutreach[3] is supporting the activities by providing training and organisational support for the medical specialists in Australia.

16.4. Uzbekistan Telemedicine Network Project

Uzbekistan, being one of the new Independent States, is confronted with numerous challenges concerning research and education particularly in health care and in information and communication technologies (ICTs). Most scientists and researchers are facing difficult a working environment due to lack of funding and severe isolation from the international scientific community.

This project is addressing these problems in a multidisciplinary way with an institutional partnership between the Department of Pathology of the University of Basel and the Republican Centre of Pathology of the Ministry of Health and the Institute for Informatics of the Academy of Sciences in Tashkent. The main purpose of the project is establishing a clinical and educational telemedicine network between Uzbekistan and Switzerland as a tool for exchange of information and knowledge.

Pathologists in Uzbekistan will be able to use the network for clinical consultations with pathologist worldwide in order to improve their skills in diagnostics and teaching and to foster their participation in the international scientific community.

A regional telemedicine network for Uzbekistan will be set up and operated by the national Institute for Informatics in Tashkent. The network will be based on the open-source telemedicine platform iPath developed at the Department of Pathology of the University of Basel. The involvement of the Institute for Informatics will ensure a long term sustainability of the telemedicine network beyond the duration of this particular project and will allow the Institute of Informatics to make the network available and useful to specialists from other medical fields (dermatology, radiology, etc.) which may equally profit from this initiative.

Finally, the telemedicine link can be used for joint educational initiatives such as distance presentations and the joint development of educational material for teaching general pathology to medical students in Uzbekistan. This will also include the application of telemedicine in interdisciplinary tumour board meetings and clinical pathological conferences where histological and cytological diagnosws can be demonstrated and explained to the physicians who are treating the patients.

The Uzbek telemedicine network will help the partners in Uzbekistan to become part of the international scientific community – in the medical field as well as in the field of ICTs. Besides the direct application of telemedicine, access to Internet and contacts with partners familiar with

---

[3]http://www.medtechoutreach.org/

international scientific publishing, it is expected that the Uzbek partners can make more efficient use of available, evidence-based health information resources (e.g. through HINARI) and eventually play an active part in international scientific publishing.

Current State

This project has been granted by the Swiss National Science Foundation within the SCOPES program. It was started in January 2006

17. Conclusions

A central part of this project consisted in the development of iPath, an Internet and email based platform for telemedicine, based on a virtual community model. The combination of web and email access to the same database has proven very efficient for enabling distance collaboration between partners with very different technical and medical backgrounds. The possibility to interact with the platform by email has ensured maximum accessibility of the platform with minimal infrastructure requirements, while the more sophisticated web interface has ensured that a large number of specialists can be easily organised into various virtual communities. The fact that iPath is very user-friendly and has been developed to accommodate a large range of telemedicine applications makes it an useful collaboration platform for groups of medical specialists within industrialised nations. This has greatly contributed to recruiting highly qualified volunteering specialists to accept occasional consultations from developing countries.

Two retrospective review studies in the field of clinical pathology revealed that delivery of highly accurate and timely diagnostic support for hospitals in developing countries is possible with relatively simple technology. Organisation of the pathologists into a "virtual institute" – a separate user group with a defined duty plan – was a key factor for reducing the turn around-time to days and often hours. This form of telemedicine can be highly recommended to organise support for a remote hospital through an international group of volunteering specialists. Sharing of a common platform among numerous projects is useful for recruiting and organising specialists and for reducing administrative overhead.

Besides improving individual diagnosis, telemedicine can be a very efficient tool for supporting geographically or professionally isolated health care providers by providing access to second opinion consultations, continuous education and exchange of knowledge. In the long term this contributes to the skills development of health care providers and will increase their capacity of handling medical problems locally. Together with the possibility to improve referral planning this is in our opinion the most significant way in which telemedicine can contribute to strengthening of health care systems at large.

Telemedicine should be understood in the context of health information access in general and there is need for a shift of focus from isolated application towards integration of telemedicine with distance learning and knowledge management, whereby the interactive and bi-directional nature of telemedicine could play an important role in improving effective knowledge transfer. Collaboration in virtual communities is a powerful way of supporting health providers in transition to evidence-based medicine.

Decentralised regional networks provide a promising approach to accommodate language barriers and cultural differences and to improve the relevancy of content and consultations with respect to local context. The technical feasibility was tested with the installation of iPath-servers in Ukraine and South Africa. In both projects the most critical factor for successful implementation and operation of telemedicine networks was not technology but organisation of work-flow and collaboration between all stakeholders. Continuous evaluation of activities regarding outcome in terms of improvement of individual health status as well as knowledge transfer should become part of such projects; and there is definitively a need for an international platform or database for the exchange of experiences with telemedicine in resource-constrained areas.

The most promising direction for further research is in our opinion an effort towards integration of different types of networks with different forms of distance collaboration and based on different existing technologies into a larger framework with a focus on continuous and sustainable transfer of knowledge in all directions and between all stakeholders of a health system.

Bibliography

[1] American Telemedicine Assocation. Definition of telemedicine. http://www.atmeda.org/news/definition.html, 1 Maj 2006.

[2] B. Aronson. Improving online access to medical information for low-income countries. *N Engl J Med*, 350(10):966–968, Mar 2004.

[3] F. Asah. *ICTs in the management of health information by medical professionals in six selected government hospitals in Yaounde, Cameroon.* University of KwaZulu Natal, 2002.

[4] M. Baba, D. Seckin, S. Kapdagli. A comparison of teledermatology using store-and-forward methodology alone, and in combination with Web camera videoconferencing. *J Telemed Telecare*, 11(7):354–360, 2005.

[5] S. Bagchi. Telemedicine in rural India. *PLoS Med*, 3(3):e82, Mar 2006.

[6] M. K. Baruah. The practice of telepathology in India. *J Postgrad Med*, 51(4):316–318, 2005.

[7] M. K. Baruah, F. G. L. Rosa. Optimal imaging in static telepathology. *Indian J Pathol Microbiol*, 45(3):367–370, Jul 2002.

[8] A. K. Bhattacharyya, J. R. Davis, B. E. Halliday, A. R. Graham, S. A. Leavitt, R. Martinez, R. A. Rivas, R. S. Weinstein. Case triage model for the practice of telepathology. *Telemed J*, 1(1):9–17, 1995.

[9] W. D. Bidgood, S. C. Horii. Introduction to the acr-nema dicom standard. *Radiographics*, 12(2):345–355, Mar 1992.

[10] W. D. Bidgood, S. C. Horii, F. W. Prior, D. E. V. Syckle. Understanding and using DICOM, the data interchange standard for biomedical imaging. *J Am Med Inform Assoc*, 4(3):199–212, 1997.

[11] M. Blunier, T. Z. T. D. D. K. Brauchli. Ict for distant medical collaboration in the ukraine swiss perinatal health project. *Ukrainian Journal of Telemedicine*, 2006.

[12] P. Bossyns, R. Abache, M. S. Abdoulaye, W. V. Lerberghe. Unaffordable or cost-effective?: introducing an emergency referral system in rural niger. *Trop Med Int Health*, 10(9):879–887, Sep 2005.

[13] P. Bossyns, R. Abache, M. S. Abdoulaye, H. Miye, A.-M. Depoorter, W. V. Lerberghe. Monitoring the referral system through benchmarking in rural niger: an evaluation of the functional relation between health centres and the district hospital. *BMC Health Serv Res*, 6:51, 2006.

[14] P. Bossyns, W. V. Lerberghe. The weakest link: competence and prestige as constraints to referral by isolated nurses in rural niger. *Hum Resour Health*, 2(1):1, Apr 2004.

[15] H. A. Brandling-Bennett, I. Kedar, D. J. Pallin, G. Jacques, G. J. Gumley, J. C. Kvedar. Delivering health care in rural Cambodia via store-and-forward telemedicine: a pilot study. *Telemed J E Health*, 11(1):56–62, Feb 2005.

[16] K. Brauchli, H. Christen, G. Haroske, W. Meyer, K. D. Kunze, M. Oberholzer. Telemicroscopy by the Internet revisited. *J Pathol*, 196(2):238–243, Feb 2002.

[17] K. Brauchli, H. Christen, P. Meyer, G. Haroske, W. Meyer, K. D. Kunze, R. Otto, M. Oberholzer. Telepathology: design of a modular system. *Anal Cell Pathol*, 21(3-4):193–199, 2000.

[18] K. Brauchli, R. Jagilly, H. Oberli, K. D. Kunze, G. Phillips, N. Hurwitz, M. Oberholzer. Telepathology on the Solomon Islands–two years' experience with a hybrid Web- and email-based telepathology system. *J Telemed Telecare*, 10 Suppl 1:14–17, 2004.

[19] K. Brauchli, M. Oberholzer. Comparison of telepathology services. *J Telemed Telecare*, 10(5):307–8: author reply 308, 2004.

[20] K. Brauchli, M. Oberholzer. The iPath telemedicine platform. *J Telemed Telecare*, 11 Suppl 2:3–7, 2005.

[21] K. Brauchli, H. Oberli, N. Hurwitz, K.-D. Kunze, G. Haroske, G. Jundt, G. Stauch, L. Banach, M. Wirdnam, M. Mihatsch, M. Oberholzer. Diagnostic telepathology: long-term experience of a single institution. *Virchows Arch*, 444(5):403–409, May 2004.

[22] K. Brauchli, D. O'mahony, L. Banach, M. Oberholzer. iPath - a Telemedicine Platform to Support Health Providers in Low Resource Settings. *Stud Health Technol Inform*, 114:11–17, 2005.

[23] bridges.org. ICT-Enabled Development Case Studies Series: The Compliance Service uses SMS technology for TB treatment. http://www.bridges.org/case_studies/137, 1 2003.

[24] D. Briscoe, C. F. Adair, L. D. Thompson, M. V. Tellado, S. B. Buckner, D. L. Rosenthal, T. J. O'Leary. Telecytologic diagnosis of breast fine needle aspiration biopsies. Intraobserver concordance. *Acta Cytol*, 44(2):175–180, 2000.

[25] T. M. Burkow, L. L. Nilsen. Success and failure in web-based medical collaboration. *J Telemed Telecare*, 11 Suppl 2:S11–S13, 2005.

[26] N. Campanella, E. Antico, L. Dini, P. Morosini. [Diagnostic efficiency on digital snapshots of the standard radiology imaging]. *Recenti Prog Med*, 95(12):566–569, Dec 2004.

[27] N. Campanella, M. Ferretti, P. Lorenzini, P. Morosini. [Accuracy of the cytopathological telediagnosis of palpable limphoadenopathies. A study design simulating the diagnostic support from remote access to health workers in the developing countries]. *Recenti Prog Med*, 96(5):226–230, May 2005.

[28] E. Caumes, V. L. Bris, C. Couzigou, A. Menard, M. Janier, A. Flahault. Dermatoses associated with travel to Burkina Faso and diagnosed by means of teledermatology. *Br J Dermatol*, 150(2):312–316, Feb 2004.

[29] Cell-Life. technology-based platform for communication, information and logistical support to manage hiv/aids in africa. http://cell-life.org.za/, 5 2006.

[30] K. Collins, I. Bowns, S. Walters. General practitioners' perceptions of asynchronous telemedicine in a randomized controlled trial of teledermatology. *J Telemed Telecare*, 10(2):94–98, 2004.

[31] K. Collins, S. Walters, I. Bowns. Patient satisfaction with teledermatology: quantitative and qualitative results from a randomized controlled trial. *J Telemed Telecare*, 10(1):29–33, 2004.

[32] S. S. Cross, T. Dennis, R. D. Start. Telepathology: current status and future prospects in diagnostic histopathology. *Histopathology*, 41(2):91–109, Aug 2002.

[33] B. L. Crowe. A review of the experience with teleradiology in australia. *J Telemed Telecare*, 7 Suppl 2:53–54, 2001.

[34] F. R. Dee, J. M. Lehman, D. Consoer, T. Leaven, M. B. Cohen. Implementation of virtual microscope slides in the annual pathobiology of cancer workshop laboratory. *Hum Pathol*, 34(5):430–436, May 2003.

[35] F. Demichelis, M. Barbareschi, S. Boi, C. Clemente, P. D. Palma, C. Eccher, S. Forti. Robotic telepathology for intraoperative remote diagnosis using a still-imaging-based system. *Am J Clin Pathol*, 116(5):744–752, Nov 2001.

[36] F. Demichelis, V. D. Mea, S. Forti, P. D. Palma, C. A. Beltrami. Digital storage of glass slides for quality assurance in histopathology and cytopathology. *J Telemed Telecare*, 8(3):138–142, 2002.

[37] G. Demiris, S. M. Speedie, L. L. Hicks. Assessment of patients' acceptance of and satisfaction with teledermatology. *J Med Syst*, 28(6):575–579, Dec 2004.

[38] S. Desai, R. Patil, R. Chinoy, A. Kothari, T. K. Ghosh, M. Chavan, A. Mohan, B. M. Nene, K. A. Dinshaw. Experience with telepathology at a tertiary cancer centre and a rural cancer hospital. *Natl Med J India*, 17(1):17–19, 2004.

[39] S. Desai, R. Patil, A. Kothari, T. Shet, S. Kane, A. Borges, R. Chinoy. Static telepathology consultation service between Tata Memorial Centre, Mumbai and Nargis Dutt Memorial Charitable Hospital, Barshi, Solapur, Maharashtra: an analysis of the first 100 cases. *Indian J Pathol Microbiol*, 47(4):480–485, Oct 2004.

[40] M. Dietel, T. N. Nguyen-Dobinsky, P. Hufnagl. The UICC Telepathology Consultation Center. International Union Against Cancer. A global approach to improving consultation for pathologists in cancer diagnosis. *Cancer*, 89(1):187–191, Jul 2000.

[41] B. E. Dunn, U. A. Almagro, H. Choi, D. L. Recla, R. S. Weinstein. Use of telepathology for routine surgical pathology review in a test bed in the Department of Veterans Affairs. *Telemed J*, 3(1):1–10, 1997.

[42] B. E. Dunn, U. A. Almagro, H. Choi, N. K. Sheth, J. S. Arnold, D. L. Recla, E. A. Krupinski, A. R. Graham, R. S. Weinstein. Dynamic-robotic telepathology: Department of Veterans Affairs feasibility study. *Hum Pathol*, 28(1):8–12, Jan 1997.

[43] E. V. Dunn, D. W. Conrath, W. G. Bloor, B. Tranquada. An evaluation of four telemedicine systems for primary care. *Health Serv Res*, 12(1):19–29, 1977.

[44] S. M. Edworthy. Telemedicine in developing countries. *BMJ*, 323(7312):524–525, Sep 2001.

[45] T. J. Eide, I. Nordrum, B. Engum, E. Rinde. [Use of telecommunications in pathology and anatomy services]. *Tidsskr Nor Laegeforen*, 111(1):17–19, Jan 1991.

[46] Electronic Radiology Laboratory Mallinckrodt Institute of Radiology. MIR DICOM central test node software.

[47] I. Ellis. The clinical champion role in the development of a successful telehealth wound care project for remote australia. *J Telemed Telecare*, 11 Suppl 2:S26–S28, 2005.

[48] O. Ferrer-Roca, R. D. D. D. Leon, F. J. de Latorre, M. Suarez-Delgado, L. D. Persia, M. Cordo. Aviation medicine: challenges for telemedicine. *J Telemed Telecare*, 8(1):1–4, 2002.

[49] H. S. Fraser, D. Jazayeri, L. Bannach, P. Szolovits, S. J. McGrath. TeleMedMail: free software to facilitate telemedicine in developing countries. *Medinfo*, 10(Pt 1):815–819, 2001.

[50] H. S. Fraser, S. J. McGrath. Information technology and telemedicine in sub-saharan Africa. *BMJ*, 321(7259):465–466, 2000.

[51] M. Fuchs. Provider attitudes toward STARPAHC: a telemedicine project on the Papago reservation. *Med Care*, 17(1):59–68, Jan 1979.

[52] K. Ganapathy. Telemedicine and neurosciences in developing countries. *Surg Neurol*, 58(6):388–394, Dec 2002.

[53] A. Geissbuhler, O. Ly, C. Lovis, J.-F. L'Haire. Telemedicine in Western Africa: lessons learned from a pilot project in Mali, perspectives and recommendations. *AMIA Annu Symp Proc*, strony 249–253, 2003.

[54] H. Gernsback. The radio doctor - maybe [cover]. *Radio News Magazine*, April 1924.

[55] J. Gershen-Cohen, A. G. Cooley. Telognosis. *Radiology*, 55(4):582–587, Oct 1950.

[56] J. Gershon-Cohen, M. B. Hermel, H. S. Read, B. Caplan, A. G. Cooley. Telognosis; three years of experience with diagnosis by telephone-transmitted roentgenograms. *J Am Med Assoc*, 148(9):731–732, Mar 1952.

[57] K. Glatz-Krieger, D. Glatz, M. Mihatsch. [Virtual microscopy: first applications.]. *Pathologe*, Aug 2005.

[58] K. Glatz-Krieger, D. Glatz, M. J. Mihatsch. Virtual slides: high-quality demand, physical limitations, and affordability. *Hum Pathol*, 34(10):968–974, Oct 2003.

[59] L. E. Graham, M. Zimmerman, D. J. Vassallo, V. Patterson, P. Swinfen, R. Swinfen, R. Wootton. Telemedicine–the way ahead for medicine in the developing world. *Trop Doct*, 33(1):36–38, Jan 2003.

[60] S. M. Gulube, S. Wynchank. Telemedicine in South Africa: success or failure? *J Telemed Telecare*, 7 Suppl 2:47–49, 2001.

[61] S. M. Gulube, S. Wynchank. The national telemedicine system in South Africa–an overview and progress report. *S Afr Med J*, 92(7):513–515, Jul 2002.

[62] M. Hadida-Hassan, S. J. Young, S. T. Peltier, M. Wong, S. Lamont, M. H. Ellisman. Web-based telemicroscopy. *J Struct Biol*, 125(2-3):235–245, 1999.

[63] D. Hailey. The need for cost-effectiveness studies in telemedicine. *J Telemed Telecare*, 11(8):379–383, 2005.

[64] B. E. Halliday, A. K. Bhattacharyya, A. R. Graham, J. R. Davis, S. A. Leavitt, R. B. Nagle, W. J. McLaughlin, R. A. Rivas, R. Martinez, E. A. Krupinski, R. S. Weinstein. Diagnostic accuracy of an international static-imaging telepathology consultation service. *Hum Pathol*, 28(1):17–21, Jan 1997.

[65] G. Haroske, K. Brauchli, M. Oberholzer. Point-to-point versus web-based telepathology in intra-operative diagnostics. *15th Int. Congr. Computer Assisted Radiology and Surgery, Berlin*, 1:27–30, 2001.

[66] H. Helin, M. Lundin, J. Lundin, P. Martikainen, T. Tammela, H. Helin, T. van der Kwast, J. Isola. Web-based virtual microscopy in teaching and standardizing Gleason grading. *Hum Pathol*, 36(4):381–386, Apr 2005.

[67] A. Y. Hira, T. T. Lopes, A. N. de Mello, V. O. Filho, M. K. Zuffo, R. de Deus Lopes. Establishment of the brazilian telehealth network for paediatric oncology. *J Telemed Telecare*, 11 Suppl 2:S51–S52, 2005.

[68] A. D. Hockey, R. Wootton, T. Casey. Trial of low-cost teledermatology in primary care. *J Telemed Telecare*, 10 Suppl 1:44–47, 2004.

[69] iPath association. website. http://ipath.ch, 2006.

[70] ITU. Report on question 14-1/2 (improving access to e-health services), 9 2005. pre-published version.

[71] P. A. Jennett, K. Andruchuk. Telehealth: 'real life' implementation issues. *Comput Methods Programs Biomed*, 64(3):169–174, Mar 2001.

[72] P. A. Jennett, M. P. Gagnon, H. K. Brandstadt. Preparing for success: readiness models for rural telehealth. *J Postgrad Med*, 51(4):279–285, 2005.

[73] P. A. Jennett, L. A. Hall, D. Hailey, A. Ohinmaa, C. Anderson, R. Thomas, B. Young, D. Lorenzetti, R. E. Scott. The socio-economic impact of telehealth: a systematic review. *J Telemed Telecare*, 9(6):311–320, 2003.

[74] O. C. Jensen, N. B. BÃžggild, S. Kristensen. Telemedical advice to long-distance passenger ferries. *J Travel Med*, 12(5):254–260, 2005.

[75] K. Johnston, C. Kennedy, I. Murdoch, P. Taylor, C. Cook. The cost-effectiveness of technology transfer using telemedicine. *Health Policy Plan*, 19(5):302–309, Sep 2004.

[76] M. Kagawa-Singer, N. Pourat. Asian American and Pacific Islander breast and cervical carcinoma screening rates and healthy people 2000 objectives. *Cancer*, 89(3):696–705, Aug 2000.

[77] R. Karlsten, B. A. Sjoqvist. Telemedicine and decision support in emergency ambulances in Uppsala. *J Telemed Telecare*, 6(1):1–7, 2000.

[78] A. Khazei, S. Jarvis-Selinger, K. Ho, A. Lee. An assessment of the telehealth needs and health-care priorities of Tanna Island: a remote, under-served and vulnerable population. *J Telemed Telecare*, 11(1):35–40, 2005.

[79] J. M. Kirigia, A. Seddoh, D. Gatwiri, L. H. K. Muthuri, J. Seddoh. E-health: determinants, opportunities, challenges and the way forward for countries in the WHO African Region. *BMC Public Health*, 5:137, 2005.

[80] E. T. Kldiashvili. Opportunities and challenges of ehealth - interconnectiviy of healthcare services. *Ukrainian Journal of Telemedicine*, 4:4–8, 2006.

[81] A. Knol, T. W. van den Akker, R. J. Damstra, J. de Haan. Teledermatology reduces the number of patient referrals to a dermatologist. *J Telemed Telecare*, 12(2):75–78, 2006.

[82] E. A. Krupinski, R. S. Weinstein, L. S. Rozek. Experience-related differences in diagnosis from medical images displayed on monitors. *Telemed J*, 2(2):101–108, 1996.

[83] E. S. Lee, I. S. Kim, J. S. Choi, B. W. Yeom, H. K. Kim, J. H. Han, M. S. Lee, A. S. Y. Leong. Accuracy and reproducibility of telecytology diagnosis of cervical smears. A tool for quality assurance programs. *Am J Clin Pathol*, 119(3):356–360, Mar 2003.

[84] S. Lee, T. J. Broderick, J. Haynes, C. Bagwell, C. R. Doarn, R. C. Merrell. The role of low-bandwidth telemedicine in surgical prescreening. *J Pediatr Surg*, 38(9):1281–1283, Sep 2003.

[85] S.-H. Lee. Virtual microscopy:applications to hematology. *Lab Hematol*, 11(1):38–45, 2005.

[86] M. Loane, R. Wootton. A review of guidelines and standards for telemedicine. *J Telemed Telecare*, 8(2):63–71, 2002.

[87] M. A. Loane, S. E. Bloomer, R. Corbett, D. J. Eedy, H. E. Gore, N. Hicks, C. Mathews, J. Paisley, K. Steele, R. Wootton. Patient cost-benefit analysis of teledermatology measured in a randomized control trial. *J Telemed Telecare*, 5 Suppl 1:S1–S3, 1999.

[88] M. A. Loane, A. Oakley, M. Rademaker, N. Bradford, P. Fleischl, P. Kerr, R. Wootton. A cost-minimization analysis of the societal costs of realtime teledermatology compared with conventional care: results from a randomized controlled trial in New Zealand. *J Telemed Telecare*, 7(4):233–238, 2001.

[89] A. M. Lopez, D. Avery, E. Krupinski, S. Lazarus, R. S. Weinstein. Increasing access to care via tele-health: the Arizona experience. *J Ambul Care Manage*, 28(1):16–23, 2005.

[90] U. Luethi, L. Risch, W. Korte, M. Bader, A. R. Huber. Telehematology: critical determinants for successful implementation. *Blood*, 103(2):486–488, Jan 2004.

[91] R. Mahendran, M. J. D. Goodfield, R. A. Sheehan-Dare. An evaluation of the role of a store-and-forward teledermatology system in skin cancer diagnosis and management. *Clin Exp Dermatol*, 30(3):209–214, May 2005.

[92] T. Mairinger. Acceptance of telepathology in daily practice. *Anal Cell Pathol*, 21(3-4):135–140, 2000.

[93] J. Marescaux, J. Leroy, M. Gagner, F. Rubino, D. Mutter, M. Vix, S. E. Butner, M. K. Smith. Transatlantic robot-assisted telesurgery. *Nature*, 413(6854):379–380, Sep 2001.

[94] W. M. Martin, S. K. Sengupta, D. P. Murthy, D. L. Barua. The spectrum of cancer in Papua New Guinea. An analysis based on the Cancer Registry 1979-1988. *Cancer*, 70(12):2942–2950, Dec 1992.

[95] V. D. Mea, P. Cataldi, S. Boi, N. Finato, P. D. Palma, C. A. Beltrami. Image selection in static telepathology through the Internet. *J Telemed Telecare*, 4 Suppl 1:20–22, 1998.

[96] V. D. Mea, P. Cataldi, S. Boi, N. Finato, P. D. Palma, C. A. Beltrami. Image sampling in static telepathology for frozen section diagnosis. *J Clin Pathol*, 52(10):761–765, Oct 1999.

[97] V. D. Mea, S. Forti, F. Puglisi, P. Bellutta, N. Finato, P. D. Palma, F. Mauri, C. A. Beltrami. Telepathology using Internet multimedia electronic mail: remote consultation on gastrointestinal pathology. *J Telemed Telecare*, 2(1):28–34, 1996.

[98] B. Mendelow. New ict technologies in support of the national arv rollout plan for hiv/aids. *2004 Eastern Cape Telehealth Conference*. University of Transkei, 2 2004.

[99] K. Mfenyana, M. Griffin, P. Yogeswaran, B. Modell, M. Modell, J. Chandia, I. Nazareth. Socio-economic inequalities as a predictor of health in south africa–the yenza cross-sectional study. *S Afr Med J*, 96(4):323–330, Apr 2006.

[100] Miniwatts Marketing Group. World internet users and population stats. http://www.internetworldstats.com/stats.htm, 3 2006.

[101] M. Mireskandari, G. Kayser, P. Hufnagl, T. Schrader, K. Kayser. Teleconsultation in diagnostic pathology: experience from Iran and Germany with the use of two European telepathology servers. *J Telemed Telecare*, 10(2):99–103, 2004.

[102] J. Monnier, R. G. Knapp, B. C. Frueh. Recent advances in telepsychiatry: an updated review. *Psychiatr Serv*, 54(12):1604–1609, Dec 2003.

[103] G. T. Moore, T. R. Willemain, R. Bonanno, W. D. Clark, A. R. Martin, R. P. Mogielnicki. Comparison of television and telephone for remote medical consultation. *N Engl J Med*, 292(14):729–732, Apr 1975.

[104] M. A. Moore, K. Tajima. Cervical cancer in the asian pacific-epidemiology, screening and treatment. *Asian Pac J Cancer Prev*, 5(4):349–361, 2004.

[105] C. Morris. Lubisi and tsilitwa cross the digital divide. http://www.iconnect-online.org/Stories/Story.import4712, 3 2002.

[106] S. Mukundan, K. Vydareny, D. J. Vassallo, S. Irving, D. Ogaoga. Trial telemedicine system for supporting medical students on elective in the developing world. *Acad Radiol*, 10(7):794–797, Jul 2003.

[107] F. G. Mullick, P. Fontelo, C. Pemble. Telemedicine and telepathology at the Armed Forces Institute of Pathology: history and current mission. *Telemed J*, 2(3):187–193, 1996.

[108] M. Musoke. Simple icts reduce maternal mortality in rural uganda. online, 1999.

[109] M. G. Musoke. Information and its value to health workers in rural Uganda: a qualitative perspective. *Health Libr Rev*, 17(4):194–202, Dec 2000.

[110] E. Neri, J. P. Thiran, D. Caramella, C. Petri, C. Bartolozzi, B. Piscaglia, B. Macq, T. Duprez, G. Cosnard, B. Maldague, J. D. Pauw. Interactive DICOM image transmission and telediagnosis over the European ATM network. *IEEE Trans Inf Technol Biomed*, 2(1):35–38, Mar 1998.

[111] I. Nordrum, T. J. Eide. Remote frozen section service in Norway. *Arch Anat Cytol Pathol*, 43(4):253–256, 1995.

[112] I. Nordrum, B. Engum, E. Rinde, A. Finseth, H. Ericsson, M. Kearney, H. Stalsberg, T. J. Eide. Remote frozen section service: a telepathology project in northern Norway. *Hum Pathol*, 22(6):514–518, Jun 1991.

[113] M. Oberholzer, H. Christen, G. Haroske, M. Helfrich, H. Oberli, G. Jundt, G. Stauch, M. Mihatsch, K. Brauchli. Modern telepathology: a distributed system with open standards. *Curr Probl Dermatol*, 32:102–114, 2003.

[114] M. Oberholzer, H. R. Fischer, H. Christen, S. Gerber, M. Bruehlmann, M. Mihatsch, M. Famos, C. Winkler, P. Fehr, L. Baechthold. Telepathology with an integrated services digital network–a new tool for image transfer in surgical pathology: a preliminary report. *Hum Pathol*, 24(10):1078–1085, Oct 1993.

[115] M. Oberholzer, H. R. Fischer, H. Christen, S. Gerber, M. Bruehlmann, M. J. Mihatsch, T. Gahm, M. Famos, C. Winkler, P. Fehr. Telepathology: frozen section diagnosis at a distance. *Virchows Arch*, 426(1):3–9, 1995.

[116] D. O'Mahony, L. Banach. Teledermatology in a rural family practice. *S A Fam Pract*, 25(6):4–8, 2002.

[117] B. B. Ong, L. M. Looi. Medico-legal aspects of histopathology practice. *Malays J Pathol*, 23(1):1–7, Jun 2001.

[118] outsource2india.com. Outsourcing teleradiology to India. http://www.outsource2india.com/services/teleradiology.asp, 13 april 2006.

[119] M. O. Oztas, E. Calikoglu, K. Baz, A. Birol, M. Onder, T. Calikoglu, M. T. Kitapci. Reliability of Web-based teledermatology consultations. *J Telemed Telecare*, 10(1):25–28, 2004.

[120] N. Paksoy, B. Montaville, S. W. McCarthy. Cancer occurrence in Vanuatu in the South Pacific, 1980-86. *Asia Pac J Public Health*, 3(3):231–236, 1989.

[121] F. Parent, Y. Coppieters, M. Parent. Information technologies, health, and "globalization": anyone excluded? *J Med Internet Res*, 3(1):E11, 2001.

[122] I. Petersen, G. Wolf, K. Roth, K. Schluens. Telepathology by the Internet. *J Pathol*, 191(1):8–14, May 2000.

[123] D. Piccolo, H. P. Soyer, W. Burgdorf, R. Talamini, K. Peris, L. Bugatti, V. Canzonieri, L. Cerroni, S. Chimenti, G. D. Rosa, G. Filosa, R. Hoffmann, I. Julis, H. Kutzner, L. Manente, C. Misciali, H. Schaeppi, M. Tanaka, W. Tyler, B. Zelger, H. Kerl. Concordance between telepathologic diagnosis and conventional histopathologic diagnosis: a multiobserver store-and-forward study on 20 skin specimens. *Arch Dermatol*, 138(1):53–58, Jan 2002.

[124] K. W. Prasse, E. A. Mahaffey, J. R. Duncan, M. F. Burrow. Accuracy of interpretation of microscopic images of cytologic, hematologic, and histologic specimens using a low-resolution desktop video conferencing system. *Telemed J*, 2(4):259–266, 1996.

[125] S. S. Raab, M. S. Zaleski, P. A. Thomas, T. H. Niemann, C. Isacson, C. S. Jensen. Telecytology: diagnostic accuracy in cervical-vaginal smears. *Am J Clin Pathol*, 105(5):599–603, May 1996.

[126] E. Rashid, O. Ishtiaq, S. Gilani, A. Zafar. Comparison of store and forward method of teledermatology with face-to-face consultation. *J Ayub Med Coll Abbottabad*, 15(2):34–36, 2003.

[127] M. Rigby. Impact of telemedicine must be defined in developing countries. *BMJ*, 324(7328):47–48, Jan 2002.

[128] N. Rogers, P. Furness, J. Rashbass. Development of a low-cost telepathology network in the UK National Health Service. *J Telemed Telecare*, 7(2):121–123, 2001.

[129] J. K. Rotich, T. J. Hannan, F. E. Smith, J. Bii, W. W. Odero, N. Vu, B. W. Mamlin, J. J. Mamlin, R. M. Einterz, W. M. Tierney. Installing and implementing a computer-based patient record system in sub-Saharan Africa: the Mosoriot Medical Record System. *J Am Med Inform Assoc*, 10(4):295–303, 2003.

[130] P. Schmid-Grendelmeier, P. Doe, N. Pakenham-Walsh. Teledermatology in sub-Saharan Africa. *Curr Probl Dermatol*, 32:233–246, 2003.

[131] P. Schmid-Grendelmeier, E. J. Masenga, A. Haeffner, G. Burg. Teledermatology as a new tool in sub-saharan Africa: an experience from Tanzania. *J Am Acad Dermatol*, 42(5 Pt 1):833–835, May 2000.

[132] J. Schneider. Telepathology at Tikur Anbessa Hospital: how telemedicine works. *Ethiop Med J*, 43(1):51–53, Jan 2005.

[133] P. Schwarzmann. Telemicroscopy. Design considerations for a key tool in telepathology. *Zentralbl Pathol*, 138(6):383–387, Dec 1992.

[134] R. E. Scott, P. Ndumbe, R. Wootton. An e-health needs assessment of medical residents in cameroon. *J Telemed Telecare*, 11 Suppl 2:S78–S80, 2005.

[135] J. Settakorn, T. Kuakpaetoon, F. J. W.-M. Leong, K. Thamprasert, K. Ichijima. Store-and-forward diagnostic telepathology of small biopsies by e-mail attachment: a feasibility pilot study with a view for future application in Thailand diagnostic pathology services. *Telemed J E Health*, 8(3):333–341, 2002.

[136] H. P. Soyer, R. Hofmann-Wellenhof, C. Massone, G. Gabler, H. Dong, F. Ozdemir, G. Argenziano. telederm.org: freely available online consultations in dermatology. *PLoS Med*, 2(4):e87, Apr 2005.

[137] B. Steffen, D. Gianom, C. Winkler, H. J. Hosch, M. Oberholzer, M. Famos. [Frozen section diagnosis using telepathology]. *Swiss Surg*, 3(1):25–29, 1997.

[138] A. Stepien, L. Banach, D. Chibowski, E. Korobowicz, L. Wronecki, E. Sobolewska. Histo- and cytopathologic remote diagnosis (telepathology). Preliminary report. *Ann Univ Mariae Curie Sklodowska [Med]*, 54:313–318, 1999.

[139] A. J. M. A. Suleiman. E-health strategies for developing countries. *IMIA Yearbook of Medical Informatics*, strony 148–156, 2005.

[140] P. Swinfen, R. Swinfen, K. Youngberry, R. Wootton. A review of the first year's experience with an automatic message-routing system for low-cost telemedicine. *J Telemed Telecare*, 9 Suppl 2:S63–S65, 2003.

[141] R. Swinfen, P. Swinfen. Low-cost telemedicine in the developing world. *J Telemed Telecare*, 8 Suppl 3:S3:63–S3:65, 2002.

[142] R. Swinfen, P. Swinfen. Low-cost telemedicine in the developing world. *J Telemed Telecare*, 8 Suppl 3(6):63–65, Dec 2002.

[143] A. Szot, F. L. Jacobson, S. Munn, D. Jazayeri, E. Nardell, D. Harrison, R. Drosten, L. Ohno-Machado, L. M. Smeaton, H. S. F. Fraser. Diagnostic accuracy of chest X-rays acquired using a digital camera for low-cost teleradiology. *Int J Med Inform*, 73(1):65–73, Feb 2004.

[144] J. Szymas, W. Papierz, M. Danilewicz. Real-time teleneuropathology for a second opinion of neurooncological cases. *Folia Neuropathol*, 38(1):43–46, 2000.

[145] H. Tanriverdi, C. S. Iacono. Diffusion of telemedicine: a knowledge barrier perspective. *Telemed J*, 5(3):223–244, 1999.

[146] C. J. Terkelsen, J. F. Lassen, B. L. Norgaard, J. C. Gerdes, S. H. Poulsen, K. Bendix, J. P. Ankersen, L. B.-H. Gotzsche, F. K. RÃžmer, T. T. Nielsen, H. R. Andersen. Reduction of treatment delay in patients with ST-elevation myocardial infarction: impact of pre-hospital diagnosis and direct referral to primary percutanous coronary intervention. *Eur Heart J*, 26(8):770–777, Apr 2005.

[147] D. J. Vassallo, F. Hoque, M. F. Roberts, V. Patterson, P. Swinfen, R. Swinfen. An evaluation of the first year's experience with a low-cost telemedicine link in Bangladesh. *J Telemed Telecare*, 7(3):125–138, 2001.

[148] D. J. Vassallo, P. Swinfen, R. Swinfen, R. Wootton. Experience with a low-cost telemedicine system in three developing countries. *J Telemed Telecare*, 7 Suppl 1:56–58, 2001.

[149] M. H. Vazir, M. A. Loane, R. Wootton. A pilot study of low-cost dynamic telepathology using the public telephone network. *J Telemed Telecare*, 4(3):168–171, 1998.

[150] D. S. Weinberg, F. A. Allaert, P. Dusserre, F. Drouot, B. Retailliau, W. R. Welch, J. Longtine, G. Brodsky, R. Folkerth, M. Doolittle. Telepathology diagnosis by means of digital still images: an international validation study. *Hum Pathol*, 27(2):111–118, Feb 1996.

[151] R. S. Weinstein. Prospects for telepathology. *Hum Pathol*, 17(5):433–434, May 1986.

[152] R. S. Weinstein. Telepathology: practicing pathology in two places at once. *Clin Lab Manage Rev*, 6(2):171–3; discussion 174–5, 1992.

[153] R. S. Weinstein. Static image telepathology in perspective. *Hum Pathol*, 27(2):99–101, Feb 1996.

[154] R. S. Weinstein. Innovations in medical imaging and virtual microscopy. *Hum Pathol*, 36(4):317–319, Apr 2005.

[155] R. S. Weinstein, K. J. Bloom, L. S. Rozek. Telepathology and the networking of pathology diagnostic services. *Arch Pathol Lab Med*, 111(7):646–652, Jul 1987.

[156] R. S. Weinstein, K. J. Bloom, L. S. Rozek. Telepathology. Long-distance diagnosis. *Am J Clin Pathol*, 91(4 Suppl 1):S39–S42, Apr 1989.

[157] R. S. Weinstein, M. R. Descour, C. Liang, G. Barker, K. M. Scott, L. Richter, E. A. Krupinski, A. K. Bhattacharyya, J. R. Davis, A. R. Graham, M. Rennels, W. C. Russum, J. F. Goodall, P. Zhou, A. G. Olszak, B. H. Williams, J. C. Wyant, P. H. Bartels. An array microscope for ultrarapid virtual slide processing and telepathology. Design, fabrication, and validation study. *Hum Pathol*, 35(11):1303–1314, Nov 2004.

[158] R. S. Weinstein, M. R. Descour, C. Liang, A. K. Bhattacharyya, A. R. Graham, J. R. Davis, K. M. Scott, L. Richter, E. A. Krupinski, J. Szymus, K. Kayser, B. E. Dunn. Telepathology overview: from concept to implementation. *Hum Pathol*, 32(12):1283–1299, Dec 2001.

[159] C. A. Wells, C. Sowter. Telepathology: a diagnostic tool for the millennium? *J Pathol*, 191(1):1–7, May 2000.

[160] J. D. Whited. Teledermatology. Current status and future directions. *Am J Clin Dermatol*, 2(2):59–64, 2001.

[161] J. D. Whited, R. P. Hall, M. E. Foy, L. E. Marbrey, S. C. Grambow, T. K. Dudley, S. K. Datta, D. L. Simel, E. Z. Oddone. Patient and clinician satisfaction with a store-and-forward teledermatology consult system. *Telemed J E Health*, 10(4):422–431, 2004.

[162] B. H. Williams, I. S. Hong, F. G. Mullick, D. R. Butler, R. F. Herring, T. J. O'Leary. Image quality issues in a static image-based telepathology consultation practice. *Hum Pathol*, 34(12):1228–1234, Dec 2003.

[163] B. H. Williams, F. G. Mullick, D. R. Butler, R. F. Herring, T. J. O'leary. Clinical evaluation of an international static image-based telepathology service. *Hum Pathol*, 32(12):1309–1317, Dec 2001.

[164] L. Williamson, N. Stoops, A. Heywood. Developing a district health information system in south africa: a social process or technical solution? *Medinfo*, 10(Pt 1):773–777, 2001.

[165] G. Wolf, I. Petersen, M. Dietel. Microscope remote control with an Internet browser. *Anal Quant Cytol Histol*, 20(2):127–132, Apr 1998.

[166] M. Woollard, K. Pitt, A. J. Hayward, N. C. Taylor. Limited benefits of ambulance telemetry in delivering early thrombolysis: a randomised controlled trial. *Emerg Med J*, 22(3):209–215, Mar 2005.

[167] R. Wootton. The possible use of telemedicine in developing countries. *J Telemed Telecare*, 3(1):23–26, 1997.

[168] R. Wootton. Telemedicine and developing countries–successful implementation will require a shared approach. *J Telemed Telecare*, 7 Suppl 1:1–6, 2001.

[169] R. Wootton. Design and implementation of an automatic message-routing system for low-cost telemedicine. *J Telemed Telecare*, 9 Suppl 1:S44–S47, 2003.

[170] R. Wootton, S. E. Bloomer, R. Corbett, D. J. Eedy, N. Hicks, H. E. Lotery, C. Mathews, J. Paisley, K. Steele, M. A. Loane. Multicentre randomised control trial comparing real time teledermatology with conventional outpatient dermatological care: societal cost-benefit analysis. *BMJ*, 320(7244):1252–1256, May 2000.

[171] R. Wootton, K. Youngberry, P. Swinfen, R. Swinfen. Prospective case review of a global e-health system for doctors in developing countries. *J Telemed Telecare*, 10 Suppl 1:94–96, 2004.

[172] R. Wootton, K. Youngberry, R. Swinfen, P. Swinfen. Referral patterns in a global store-and-forward telemedicine system. *J Telemed Telecare*, 11 Suppl 2:S100–S103, 2005.

[173] World Health Organisation. ehealth. World Health Assembly no. 58 conf. paper 21, April 2005.

[174] World Health Organisation. Knowledge management strategy, who/eip/kms/2005.1. WHO/EIP/KMS/2005.1, 2005.

[175] World Summit on the Information Society. Plan of action. Document WSIS-03/GENEVA/DOC/5-E, 12 2003.

[176] D. Wright. The International Telecommunication Union's report on Telemedicine and Developing Countries. *J Telemed Telecare*, 4 Suppl 1:75–79, 1998.

[177] D. Wright, L. Androuchko. Telemedicine and developing countries. *J Telemed Telecare*, 2(2):63–70, 1996.

[178] K. Yamashiro, N. Kawamura, S. Matsubayashi, K. Dota, H. Suzuki, H. Mizushima, F. Wakao, N. Azumi. Telecytology in Hokkaido Island, Japan: results of primary telecytodiagnosis of routine cases. *Cytopathology*, 15(4):221–227, Aug 2004.

[179] G. Yamey. The professor of "telepreventive medicine". *BMJ*, 328(7449):1158, May 2004.

[180] P. Yellowlees. Successful development of telemedicine systems–seven core principles. *J Telemed Telecare*, 3(4):215–22; discussion 222–3, 1997.

[181] R. I. Zbar, L. R. Otake, M. J. Miller, J. A. Persing, D. L. Dingman. Web-based medicine as a means to establish centers of surgical excellence in the developing world. *Plast Reconstr Surg*, 108(2):460–465, Aug 2001.

[182] M. Zolfo, L. Arnould, V. Huyst, L. Lynen. Telemedicine for HIV/AIDS Care in Low Resource Settings. *Stud Health Technol Inform*, 114:18–22, 2005.

Die VDM Verlagsservicegesellschaft sucht für wissenschaftliche Verlage abgeschlossene und herausragende

Dissertationen, Habilitationen, Diplomarbeiten, Master Theses, Magisterarbeiten usw.

für die kostenlose Publikation als Fachbuch.

Sie verfügen über eine Arbeit, die hohen inhaltlichen und formalen Ansprüchen genügt, und haben Interesse an einer honorarvergüteten Publikation?

Dann senden Sie bitte erste Informationen über sich und Ihre Arbeit per Email an *info@vdm-vsg.de*.

Sie erhalten kurzfristig unser Feedback!

VDM Verlagsservicegesellschaft mbH
Dudweiler Landstr. 99
D - 66123 Saarbrücken
www.vdm-vsg.de

Telefon +49 681 3720 174
Fax   +49 681 3720 1749

Die VDM Verlagsservicegesellschaft mbH vertritt

Printed by Books on Demand GmbH, Norderstedt / Germany